Introduction to Modern Optics
for Students in Engineering and Applied Science

Stephen Arnold and Kaitlynn Snyder

NYU Tandon School of Engineering

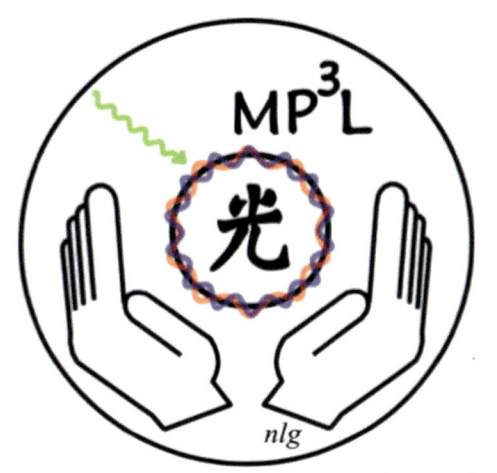

MicroParticle PhotoPhysics Lab
Last revised on 1/31/2024

Published August, 2018 by MicroParticle PhotoPhysics Lab
Brooklyn, New York 11201

Copyright © 2018 by Stephen Arnold and Kaitlynn Snyder

ISBN 978-0692172223

Preface

The following is a text put together by Kaitlynn Snyder and myself from a one-semester course I taught to students under course number PH 3474 at the NYU Tandon (Polytechnic) School of Engineering over the last two years. The course, which met for 4 hours each week, was also cross-listed for electrical engineering and chemical engineering students under the course numbers EE 3474 and CBE 3474. Unlike other texts in Modern Optics this text is intended to be used by students in both engineering and applied science at a junior or senior level, and to support specialized interdisciplinary courses given at a graduate level, such as Bio-Optics. By introducing it early in the junior year students with interest arrive fresh from their introductory physics courses. The course emphasizes fundamentals starting with Maxwell's equations, which is where the introductory physics sequence ends, and applies these fundamentals to current interests in science and technology. Appropriate to the level of the course, the mathematics represents Maxwell's Equations in their integral form. Where more advanced math was added (e.g. Fourier Transform), the students were introduced to this as if it were taught in an applied math course (see Appendices).

There are also take-home laboratory experiments for which the students acquire a kit known as Pocket Optics PO-1. With this kit, 10 experiments are assigned to support the concepts taught in the course. One of these involves turning a Smart phone into a microscope.

Applications: Some of the applications discussed are Optical Tweezers, Holographic Diffraction Grating, Demystifying the structure of DNA from Rosalind Franklin's X-ray diffraction image (Photo 51), Fourier Transform Infrared Spectroscopy (FTIR), nano-plasmonics, Fabry-Perot resonator, Whispering Gallery Mode sensor, LASER, Confocal microscope, and Super high-resolution microscopy (STED).

Front Cover: Image from an American Scientist article published in October 2001 [S. Arnold, *American Scientist* **89** 414-421 (2001)]. It shows a micro-sphere sitting on an optical fiber, and immersed in water. The person viewing from above notices that light scattering from the sphere "blinks" as the laser feeding the fiber is tuned. This discovery was first made in 1994 at my laboratory (MicroParticle PhotoPhysics Lab, MP3L) at the Polytechnic University (currently, NYU Tandon School of Engineering) and is a basis for what has become known as the Whispering Gallery Mode biosensor (WGM biosensor). The WGM biosensor works because the frequency of each resonance peak shifts upon the binding of biomolecules to the micro-sphere's surface. Extreme sensitivity to surface layer thickness was predicted in the 2001 article. The WGM biosensor is introduced in Section 6.4 just after the Fabry-Perot resonator and has evolved into the world's most sensitive biosensor by the addition of a nano-plasmonic receptor. Although the theory for enhanced sensitivity is outside the scope of this text, nano-plasmonics is discussed briefly in Section 6.1. The Whispering Gallery Mode biosensor is one of the special topics analyzed in this book.

Stephen Arnold, Fellow Optical Society (OSA),
August 10, 2018

Table of Contents

Introduction   viii

Chapter One – The Physics of Electromagnetic Waves; Maxwell's Synthesis

     1.1 – Maxwell's Equations   1
     1.2 – The Wave Equation   4
     1.3 – Solutions to the Wave Equation   6
     1.4 – The Wave Equation from Maxwell's Equations   10
     1.5 – Measuring the Speed of Light   15
     1.6 – Electromagnetic Waves in Three Dimensions   18
     1.7 – The Electromagnetic Spectrum   21
     1.8 – Energy Density and Intensity of Light   22
     1.9 – Photons, Momentum of Light, and Forces Exerted by Light   27
     1.10 – Optical Levitation and Optical Tweezers   29
     1.11 – Chapter One Exercises, including Experiment 1   37

Chapter Two – Diffraction

     2.1 – Superposition of Waves   43
     2.2 – Superposition in the Complex Plane   45
     2.3 – Constructing a Holographic Diffraction Grating   46
     2.4 – Diffraction from a Pinhole   50
     2.5 – Diffraction and the Fourier Transform   51
     2.6 – Dirac Delta Function   56
     2.7 – Fraunhofer Diffraction in Multiple Dimensions   59
     2.8 – Grating Equation at Oblique Incidence   66
     2.9 – X-ray Diffraction and the Structure of DNA   67
     2.10 – Chapter Two Exercises, including Experiments 2 & 3   75

Chapter Three – The Classical Atom and Dielectric Theory

     3.1 – Refractive Index, Snell's Law, Dispersion,
             and Chromatic Aberration   81
     3.2 – The Lorentz Atom   83
     3.3 – Radiation Force and Absorption Cross Section   87
     3.4 – The Lorentz Atom Natural Frequency and Size   89
     3.5 – Dielectric Theory, Refractive Index, and
             Chromatic Aberration   91
     3.6 – Phase Velocity and Group Velocity   98
     3.7 – Beer-Lambert Law: Light Attenuation in Solution   102
     3.8 – Chapter Three Exercises, including Experiment 4   103

Chapter Four – Polarization of Light

4.1 – Edwin Land and the Motivation for Polarized Spectacles	107
4.2 – Generation of Polarized Light	108
4.3 – Linear Dichroism and the Origin of Polarizing Spectacles	111
4.4 – Birefringence and the Generation of Circularly Polarized Light	114
4.5 – Polarization and Jones' Calculus	117
4.6 – Chapter Four Exercises, including Experiment 5	122

Chapter Five – The Theory of Reflection and Refraction at an Interface

5.1 – Reflection and Transmission Coefficients	124
5.2 – Snell's Law and Brewster's Angle	128
5.3 – Total Internal Reflection and Evanescent Field	132
5.4 – Light Interactions with Metals	134
5.5 – Chapter Five Exercises, including Experiment 6	136

Chapter Six – Optical Resonators

6.1 – Resonators as tuners: From macro-AM tuners to Plasmonic PetaHz-nano-tuners	142
6.2 – The Fabry-Perot Etalon	146
6.3 – Frequency and Time Domain as Conjugates	153
6.4 – Whispering Gallery Mode Resonator and Sensor	158
6.5 – Chapter Six Exercises	161

Chapter Seven – LASER

7.1 – What would happen if a Fabry-Perot Etalon had positive gain?	163
7.2 – Stimulated emission	164
7.3 – Original Laser and other types	168
7.4 – Chapter Seven Exercises, including Experiment 7	172

Chapter Eight – Spectroscopy

8.1 – Grating Spectrometer	176
8.2 – Fourier Transform Infrared Spectroscopy (FTIR) and the Michelson Interferometer	180
8.3 – Chapter Eight Exercises	185

Chapter Nine – Geometrical Optics

9.1 – Fermat's Principle of Least Time	188
9.2 – Constructing a Lens	196
9.3 – Optical Power of a Two-Lens System	200

9.4 – Matrix Optics   201
9.5 – Chapter Nine Exercises, including Experiment 8   205

Chapter Ten – Optical Instruments

10.1 – The Human Eye   210
10.2 – The Magnifying Glass   215
10.3 – The Compound Microscope   218
10.4 – Abbe's Diffraction Limit   219
10.5 – Total Internal Reflection Fluorescence Microscopy
       (TIRFM)   220
10.6 – The Confocal Microscope   221
10.7 – Breaking the Limit with STED Microscopy   225
10.8 – Chapter Ten Exercises, including Experiment 9   230

Appendix

A.1 – Sums and Series   234
A.2 – Complex Representation of Electromagnetic Waves   237
A.3 – Time Averages   241
A.4 – Fourier Transforms   243
A.5 – Vector Calculus for Expressing Maxwell's Eqs. as
       Differential Eqs.   246
A.6 – Take-home Experiments   249

References   250
Answers to Selected Exercises   253
Index   251

Introduction

More than any other of our senses we are aware of light soon out of the womb. Our religious training introduces creation by referring to a void that is soon illuminated by light. Through Maxwell's synthesis the description of light is as an electromagnetic phenomena, which covers the range from radio waves to gamma rays. All of these will be referred to as "light" in this text, and what we see will be distinguished as "visible light". This itself is a modern point of view.

A few years back I (S.A.) decided that there had to be an early path for engineering students interested in the subject not to stop where their introductory physics sequence stops, with a cursory introduction to Maxwell's masterful equations (1865). The notes from my lectures on the subject to 1st semester juniors through senior students were assembled into this text by K. Snyder, who at the time had graduated with a B.S. in Applied Physics. Discussions with her modified the text, making the material more accessible to students at this level.

Optics is the science of light, and the way this science is used to design instruments for measurement and manipulation of material objects. I guess you could call this text an Introduction to Applied Optics, however the emphasis is on basics, with the applications used to point out their utility.

The applications that have been chosen are indeed modern. Just a few that are not typically in textbooks on optics: (1) the 2018 Nobel prize to Arthur Ashkin for Optical Tweezers (2018), (2) the use of x-ray diffraction by DNA for defining its structure (1953) through the great experimental skill of Rosalind Franklin, (3) the Nobel prize awarded to Stefan Hell (2014) for applying basic ideas of Albert Einstein in order to innovate a means for super resolution microscopy, and (4) my own favorite - the Whispering Gallery Mode Sensor.

Scan to acquire "Pocket Optics" kit

Near the end of each chapter are homework exercises that are both computational and experimental. Students attending the NYU Tandon School of Engineering are provided with a **take-home kit** known as "**Pocket Optics**", that includes all of the parts for 9 experiments along with a base plate for mounting them (shown below). We are indebted to David Keng for designing this kit based on mutual discussions, and for making it available for sale on Amazon.

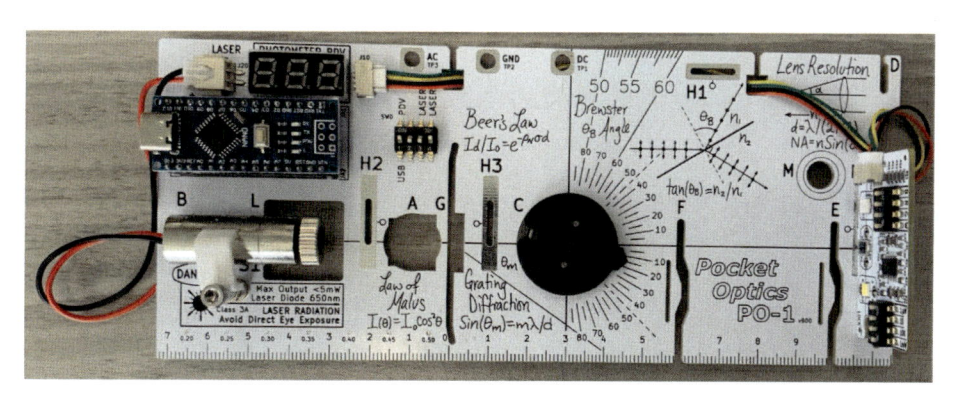

Scan for videos of the "Pocket Optics" kit in action

We start in Chapter One where introductory physics leaves off – with Maxwell's synthesis.

Chapter One – The Physics of Electromagnetic Waves; Maxwell's Synthesis

1.1 – Maxwell's Equations

One of the great accomplishments in physics took place in the 19th century at the hand of James Clerk Maxwell (Maxwell, 1865). In the 1860s Maxwell was able to show that electricity and magnetism are not separate phenomena but are intimately intertwined. At this time, it was also discovered that light is a form of electromagnetic radiation and can be described by the laws of electricity and magnetism. And so, in the 19th century there was a beautiful synthesis of three phenomena, electricity, magnetism, and light, into one theory that is entirely embodied in four elegant equations[1].

The first of these equations is Gauss's law,

$$\oint_S \underline{E} \cdot \hat{n} \, dA = \frac{Q_{enc}}{\varepsilon_0}. \tag{1.1}$$

Gauss's law states that the total flux of the electric field $\underline{E}$ through any closed surface is proportional to the enclosed charge Q_{enc} and inversely proportional to the permittivity of the medium containing the charge.[2] The differential flux is calculated as a projection of the field onto the local unit normal $\hat{n}$ pointing out from the surface times the differential area, dA. In Eq. (1.1)-(1.4) we list Maxwell's equations as they appear in vacuum. This is indicated by the zero subscript in μ_0 and ε_0. Maxwell's equations are valid in other media as well. In a homogeneous medium μ_0 will be replaced by μ and ε_0 by ε, where μ the

---

[1] This statement is a bit of an exaggeration. In the 20th century the laws of electricity, magnetism, and light had to be updated for strange new observations related to quantum phenomena. The new theory, quantum electrodynamics (QED) is also quite elegant and can be learned about in R.P. Feynman, *QED: The Strange Theory of Light and Matter*, (Princeton University Press, 2014).
[2] I will underline vector quantities throughout the text, and put a hat over unit vectors (e.g. $\hat{n}$).

permeability, and ε the permittivity, are intrinsic properties of a medium relating to how it interacts with the electric and magnetic fields[3].

The second of the equations is often called Gauss's law for magnetism,

$$\oint_S \underline{B} \cdot \hat{n} \, dA = 0. \tag{1.2}$$

Eq. (1.2) is a statement that the total flux of the magnetic field $\underline{B}$ through a closed surface is always zero. This means we cannot isolate the north and south poles of a magnet; there are no magnetic monopoles.

The third equation is Faraday's Law;

$$\oint_\ell \underline{E} \cdot d\underline{s} = -\frac{d}{dt} \int \underline{B} \cdot \hat{n} \, dA. \tag{1.3}$$

This law says that the time rate change in a magnetic flux through a closed loop will induce a circuitous electric field around the loop, or electromotive force (emf) on a closed current loop. If a magnetic field is constant, then moving the current loop in such a way that the magnetic flux thru the coil is changed produces the same effect. Converting mechanical energy to electrical energy using the principle described in Faraday's law can produce hydroelectric power. Water is used to spin powerful magnets around a loop of wire to induce a current and produce electrical power.

The last of the equations is the Maxwell-Ampère equation,

$$\oint_\ell \underline{B} \cdot d\underline{s} = \mu_0 \int \underline{J} \cdot \hat{n} \, dA + \mu_0 \varepsilon_0 \frac{d}{dt} \int \underline{E} \cdot \hat{n} \, dA. \tag{1.4}$$

This law says that a circulating magnetic field is produced around a loop, by the flux of electrical current density $\underline{J}$ (i.e. electrical current) and time derivative of electric flux through the loop. The first term on the right hand side of the equality results from an experimental discovery by Ampère and describes the magnetic field produced by an electric current. This is sufficient for currents of charge, but the magnetic field is also generated by time varying electric fields that lead to what is known as displacement current. It was Maxwell who generalized the

---

[3] We will elaborate on this a bit in later chapters, however an in-depth treatment can be found in Griffiths, *Introduction to Electrodynamics* (3rd ed.), (Prentice Hall, 1999). p. 180, 274.

equation by adding the second term, which says that a changing electric field produces a magnetic field. As a child I remember coiling a copper wire around a nail and attaching a battery to the ends of the wire to create a magnetic field; I could effect the direction of a compass placed near either end. After maintaining the current for a short time I removed the nail and found it was magnetized! At the time I didn't know the theory but was quite excited by the discovery. It turns out that there are a huge number of applications related to Maxwell's equations. Transformers, inductors, generators and solenoids are a few of the devices that use these principles.

What we have not stated explicitly is how the electric and magenetic fields act to generate a force on an individual charge q. So long as the charge is stationary the force on it is $q\underline{E}$, which provides the definition of the electrostatic field. However, a moving charge with velocity $\underline{v}$ is subject to an additional force by the magnetic field $\underline{B}$, $q\underline{v} \times \underline{B}$. Together, the overall force on a charge by both the electric and magnetic fields,

$$\underline{F} = q(\underline{E} + \underline{v} \times \underline{B}), \qquad (1.5)$$

which is commonly known as the Lorentz force.

Maxwell's equations [Eqs. (1.1)-(1.4)] as we have listed them are in integral form. Although we will use these throughout this text, Maxwell's equations can be written in the form of differential equations. This description may be less familiar to you, however the underlying phenomena are the same, and often problems solve in fewer steps. The equations use the vector *Del* operator

$$\nabla = \hat{x}\frac{\partial}{\partial x} + \hat{y}\frac{\partial}{\partial y} + \hat{z}\frac{\partial}{\partial z}; \qquad (1.6)$$

see Appendix A.5. In this form Maxwell's equations in free-space corresponding to Eq. (1.1) - Eq. (1.4) are

$$\nabla \cdot \underline{E} = \frac{\rho}{\varepsilon_0}, \qquad (1.7)$$

$$\nabla \cdot \underline{B} = 0, \qquad (1.8)$$

$$\nabla \times \underline{E} = -\frac{\partial \underline{B}}{\partial t}, \text{ and} \qquad (1.9)$$

$$\nabla \times \underline{B} = \mu_0 \underline{J} + \mu_0 \varepsilon_0 \frac{\partial \underline{E}}{\partial t}, \qquad (1.10)$$

respectively, where ρ and $\underline{J}$ are charge density and current density, respectively. It is important to understand that these equations are no more complete physically than the integral equations [i.e. Eq. (1.1) - Eq. (1.4)].

Although not central to our study, it is important to recognize that there are limits in which Maxwell's equations do not offer a complete explanation of optical phenomena. In these cases the laws of quantum physics govern light. We will do a brief overview of some quantum physical effects later in the text, but for the time being we will consider light classically, in limits for which Maxwell's equations offer an accurate description.

Next, we would like to see how Maxwell's equations come together to predict electromagnetic waves. To do that, we have to appreciate the structure of the wave equation, which we will introduce by examining the motion of a string.

1.2 – The Wave Equation

Imagine holding the end of a rope while it is attached to a wall, a great distance away and under tension. By moving the end of the rope up and down a wave with vertical displacement (a.k.a transverse) is seen to propagate toward the wall. In what follows we will derive the dynamical equation that describes this wave.

Let's begin by examining a force diagram of a small segment of the rope as it moves in a transverse wave.

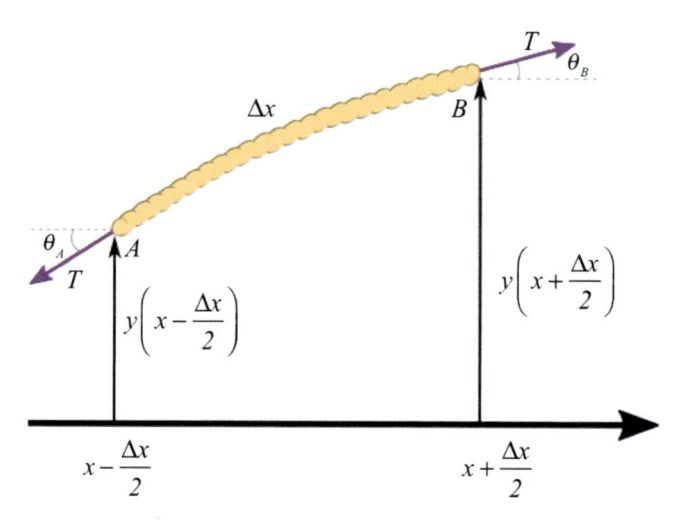

Figure 1.1 – The forces on a small segment of rope.

Fig. 1.1 depicts a snapshot of a piece of a moving wave; it is frozen in time. A careful analysis will allow us to describe the string's motion at subsequent times provided that the system remains undisturbed and we treat the string as ideal, meaning we ignore losses of energy associated with the string's structure. To begin let's note that the x-component of the string's position remains stationary and only motion in the y-direction is considered. We ignore gravity and only consider the tension forces at the ends of the segment; points A and B which are separated by a distance Δx. By examining the angles to tangent lines in Fig. 1.1 at both endpoints, we recognize that their tangents can be described by first partial derivatives of the rope's shape $y(x)$ with respect to x,

$$\tan(\theta_A) = \left.\frac{\partial y}{\partial x}\right|_{x-\frac{\Delta x}{2}} \quad \text{and} \tag{1.11}$$

$$\tan(\theta_B) = \left.\frac{\partial y}{\partial x}\right|_{x+\frac{\Delta x}{2}}. \tag{1.12}$$

Since the segment of rope does not move in the x-direction, we will only consider forces components that move it up and down; the y-components of the tensions at points A and B are

$$T_{Ay} = -T \sin(\theta_A), \text{ and} \tag{1.13}$$

$$T_{By} = T \sin(\theta_B). \tag{1.14}$$

The negative sign in Eq. (1.13) arises from recognizing that the tension on the left is pulling the rope downward; the vertical component of $\underline{T}$ at point A is in the negative y-direction, as seen in the diagram. For small angles, $\sin(\theta) \simeq \tan(\theta)$, so we can rewrite Eq. (1.13) and Eq. (1.14) in terms of the partial derivatives in Eq. (1.11) and (1.12),[4]

$$T_{Ay} \simeq -T \left.\frac{\partial y}{\partial x}\right|_{x-\frac{\Delta x}{2}} \quad \text{and} \tag{1.15}$$

---

[4] Setting $\sin(\theta) \simeq \tan(\theta)$ is equivalent to neglecting 2nd order terms such as $(\partial y/\partial x)^2$. Such terms make a negligible contribution for a taut string; $(\partial y/\partial x)^2 \ll (\partial y/\partial x)$.

$$T_{By} \simeq T \frac{\partial y}{\partial x}\bigg|_{x+\frac{\Delta x}{2}}. \qquad (1.16)$$

This gives us the vertical force component at the ends of the string segment, from which we can write the dynamical equation of motion for the segment using Newton's second law, $\underline{F} = m\underline{a}$. The mass m will be replaced with $\mu \Delta x$, where μ is the mass per unit length of the string and Δx is length of the segment in the taut string limit.[4] The acceleration in the y-direction at a fixed x requires a second partial derivative of y with respect to time; $a_y = \partial^2 y / \partial t^2$. The y-component of Newton's 2nd Law is

$$\mu \Delta x \frac{\partial^2 y}{\partial t^2}\bigg|_x = T_{By} + T_{Ay} = T\frac{\partial y}{\partial x}\bigg|_{x+\frac{\Delta x}{2}} - T\frac{\partial y}{\partial x}\bigg|_{x-\frac{\Delta x}{2}}. \qquad (1.17)$$

We can simplify this expression by dividing both sides of the equation by T and Δx,

$$\frac{\mu}{T} \frac{\partial^2 y}{\partial t^2}\bigg|_x = \frac{\frac{\partial y}{\partial x}\bigg|_{x+\frac{\Delta x}{2}} - \frac{\partial y}{\partial x}\bigg|_{x-\frac{\Delta x}{2}}}{\Delta x}. \qquad (1.18)$$

If you examine Eq. (1.18) you will note that in the limit in which the string segment has an infinitesimal length, the right hand side is just the second derivative of the displacement $y(x,t)$, with respect to x. The form may be a bit different from what you are used to so be sure to convince yourself it is a derivative before moving on. Finally, we have the equation of motion for the string

$$\frac{\mu}{T} \frac{\partial^2 y}{\partial t^2} = \frac{\partial^2 y}{\partial x^2}. \qquad (1.19)$$

Eq. (1.19) is known as the wave equation.

1.3 – Solutions to the Wave Equation

Now that we have the dynamical equation for the string the next step is to explore solutions to this equation. Rather than deriving a general solution by the method of separation of variables we will take a heuristic approach and guess at a solution based on physical observation. We know that if a horizontal rope is attached to a vertical wall and we shake the free end up and down with our hand sinusoidally with time, then a sinusoidal shaped spatial wave will propagate away from our hand. Such a wave is known as a traveling wave. To generate such a wave mathematically we imagine a coordinate system moving with the wave; the wave is frozen in this coordinate system, as shown in Fig. 1.2. The frozen wave in this figure has the equation

$$y(x') = A\sin\left(\frac{2\pi}{\lambda}x'\right) \qquad (1.20)$$

where λ is the separation between crests, known as the wavelength, and A is the largest y-displacement, known as the amplitude.

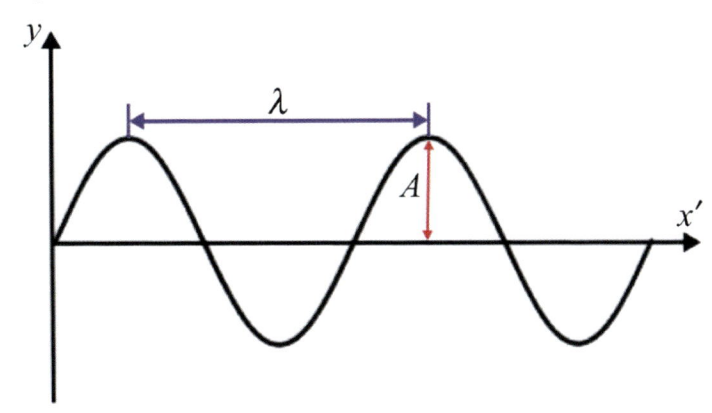

Figure 1.2 – A sine wave with amplitude A and period λ.

Now we want to make the frozen wave travel to the right. Fig. 1.3 shows how to accomplish this. It involves an observer watching from another coordinate system S while the S' system containing the "frozen" wave is translated at velocity v.

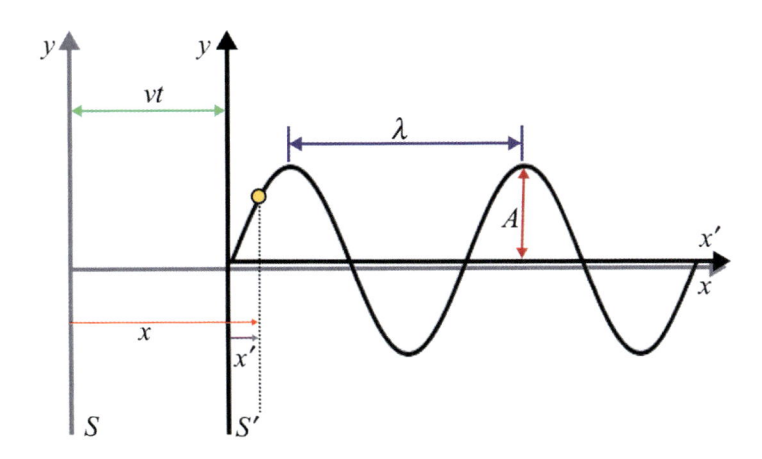

Figure 1.3 – The sine wave from Fig. 1.2 now moves along the x-axis with velocity v.

Each point at x' on the wave in S' is tracked by another coordinate x in S for which the observer is at rest. This means that

$$x = x' + vt. \tag{1.21}$$

To describe the wave in a reference frame at rest relative to the traveling wave we simply have to substitute Eq. (1.21) into Eq. (1.20). This gives a wave of the form

$$y(x,t) = A\sin\left(\frac{2\pi}{\lambda}(x-vt)\right). \tag{1.22}$$

This is the equation for a travelling wave. Next, we need to test whether it is a solution to the wave equation, Eq. (1.19).

To test this we take the second partial derivatives of $y(x,t)$ with respect to t and x,

$$\frac{\partial^2 y}{\partial t^2} = -A\left(\frac{2\pi v}{\lambda}\right)^2 \sin\left(\frac{2\pi}{\lambda}(x-vt)\right), \text{ and} \tag{1.23}$$

$$\frac{\partial^2 y}{\partial x^2} = -A\left(\frac{2\pi}{\lambda}\right)^2 \sin\left(\frac{2\pi}{\lambda}(x-vt)\right) \tag{1.24}$$

Comparing these last two equations we see that

$$\frac{\partial^2 y}{\partial t^2} = v^2 \frac{\partial^2 y}{\partial x^2}, \tag{1.25}$$

which is the same as the wave equation [Eq. (1.19)] so long as

$$v^2 = \frac{T}{\mu}. \tag{1.26}$$

So the velocity of the waves on a rope increases with increasing tension and decreases with the linear mass density, however more importantly, Eq. (1.22) proves to be a solution to the wave equation. In fact, any function of $x - vt$ is a bona fide solution to the wave equation. For example,

$$y(x,t) = e^{-a(x-vt)^2} \tag{1.27}$$

should be such a physical wave, a supposition that you should verify by testing to see if it is a solution to Eq. (1.19). If v is positive the wave is moving to the right, if v is negative the wave is moving to the left. A 2nd order partial differential equation requires two solutions, and those solutions to Eq. (1.19) are counter propagating waves. So the general solution is a superposition of a wave traveling in the $+x$ direction $f(x-vt)$, and a wave travelling in the minus $-x$ direction, $g(x+vt)$.

The solution to the wave equation can be written in terms of angular frequencies. To get a feeling for this we need to relate the velocity of a wave to its frequency and wavelength. Suppose you are standing to the right of the moving wave in Fig. 1.3. The wave hits you with one crest followed by another. The time between these impacts P_t is known as the period. If 0.1 seconds elapses between crests then you will be hit with 10 crests per second. That is the temporal frequency of the wave f. It is obtained by taking one over the period;

$$f = \frac{1}{P_t}. \qquad (1.28)$$

The wave velocity is found by dividing the distance between crests by the time between these impacts

$$v = \frac{\lambda}{P_t} = f\lambda. \qquad (1.29)$$

The frequency is often referred to in an angular form. The angular-temporal frequency, or simply angular frequency, ω is

$$\omega = \frac{2\pi}{P_t}. \qquad (1.30)$$

Another frequency often defined is the angular-spatial frequency, or simply spatial frequency, k. This spatial frequency is often known as the "propagation constant". It is 2π divided by the spatial period λ;

$$k = \frac{2\pi}{\lambda}. \qquad (1.31)$$

In terms of these "angular frequencies" the velocity in Eq. (1.29) is

$$v = \frac{\omega}{k}, \qquad (1.32)$$

and the sinusoidal wave solution is

$$y(x,t) = A\sin(kx - \omega t). \qquad (1.33)$$

The argument of the sinusoidal function is known as the phase, and Eq. (1.32) is often called the "phase velocity", v_p, because it corresponds to the velocity of a

9

point having constant phase; with the phase $kx - \omega t =$ constant; $kdx - \omega dt = 0$ and $v_p = dx/dt = \omega/k$.

1.4 – The Wave Equation from Maxwell's Equations

The wave we would like to describe now travels without needing a medium and is generated by accelerating charges. Imagine an infinite sheet of oscillating current that is infinitesimally thin. A small section of the sheet is shown in Fig. 1.4. We will show through the application of Maxwell's equations that it will emit electromagnetic waves. This oscillating current replaces the hand we used to stimulate waves on the rope, and the electric and magnetic fields generated in front and in back of the current sheet will end up following wave equations. We have drawn three loops on this figure that will aid in describing the physics of this transmitting antenna as we move forward.

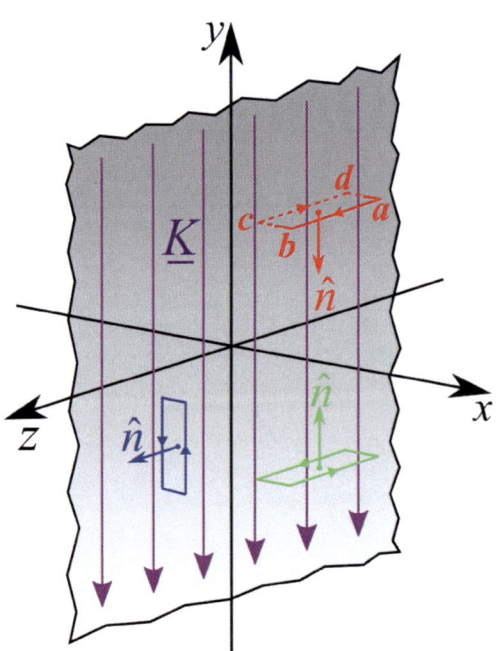

Figure 1.4 – Current carrying sheet in the yz plane.

At the instant shown in Fig.1.4, which we will mark as time $t = 0$, the current per unit width is in the negative y-direction, with y-component $-K_0$, where K_0 is the amplitude of the oscillating current. The magnetic field at the surface is obtained from the Maxwell-Ampere equation [Eq. (1.4)],

$$\oint_\ell \underline{B} \cdot d\underline{s} = \mu_0 \int \underline{K}_0 \cdot \hat{n} \, dz + \mu_0 \varepsilon_0 \frac{d}{dt} \int \underline{E} \cdot \hat{n} dA, \qquad (1.34)$$

where we have replaced $\underline{J}dA$ with $\underline{K}dz$. First we evaluate the path integral of the magnetic field over the red loop in Fig. 1.4. This loop cuts through the sheet with path a→b along +z on the front side, and c→d along −z on the back side. Since the sheet is infinitesimally thin, the length of the sides penetrating the current sheet (i.e bc and da) can be reduced toward zero, thereby excluding any electric flux. In addition, since the magnetic field must circulate around the current, its vector must lie parallel to the xz plane. A component in the x-direction is excluded by Gauss's law of magnetism, Eq. (1.2). This can be shown through a simple symmetry argument. If the magnetic field at the surface points outward along +x on the front side of the surface, symmetry requires it must point outward along −x on the back side. That leads to a net magnetic flux from the surface, which would require a net magnetic charge in violation of Eq. (1.2). Therefore, the magnetic field can only be along the z-direction. Now, the path integral, a→b→c→d→a, only has two segments with non-zero contributions; a→b and c→d. The Maxwell-Ampere equation yields

$$B_z(0^+)w - B_z(0^-)w = \mu_0 K_0 w, \qquad (1.35)$$

where w is the path length in the front of the plane, and $B_z(0^+)$ and $B_z(0^-)$ are the magnetic fields on the front and back side of the sheet, respectively. From symmetry, if the field is directed toward +z in front of the plane it must be directed toward −z behind the plane, consequently $B_z(0^-) = -B_z(0^+)$. That generates a magnetic field on the positive side of the surface in the positive z-direction with magnitude $B_z(0^+) = \mu_0 K_0/2$. With increasing time the oscillating current generates a magnetic field on the front surface

$$B_z(0^+, t) = (\mu_0/2) K_0 \cos(\omega t). \qquad (1.36)$$

The magnetic field will keep its z-polarization as it extends into space in front and in back of the sheet, although the magnetic field on the back surface is reversed. Due to the infinite size of the sheet in the y-z plane, the magnetic field is invariant with respect to y and z, and as a consequence can only depend on x and t. The sinusoidal varying magnetic field will lead to a time varying electric field from Faraday's law [Eq. (1.3)]

$$\oint_{loop} \underline{E} \cdot d\underline{s} = -\frac{d}{dt} \int \underline{B} \cdot \hat{n} dA. \qquad (1.3)$$

In order to understand the fields in the space just in front of the current sheet, we will evaluate the path integral in Eq. (1.3) over the blue loop just in front of the current sheet (Fig. 1.4) with its center at x, and having width Δx, and height L.

We start at the lower right hand corner of this blue rectangle in Fig. 1.4 and add up the projection of the field on the line elements in a counter-clockwise direction. Since the sheet is charge neutral there is no x-component of the electric field, so the parts of the path integral perpendicular to the sheet make no contribution. On the right of Eq. (1.3) we need to multiply the magnetic field in the z-direction times the loop area. This is the magnetic flux for which we must take the temporal derivative. Together

$$E_y\big|_{x+\frac{\Delta x}{2}} L - E_y\big|_{x-\frac{\Delta x}{2}} L = -\frac{d}{dt}\int \underline{B}\cdot \hat{n}dA = -\frac{\partial}{\partial t} B_z L \Delta x. \qquad (1.37)$$

Simplifying Eq. (1.37) gives

$$E_y\left(x+\frac{\Delta x}{2}\right) - E_y\left(x-\frac{\Delta x}{2}\right) = -\frac{\partial}{\partial t} B_z \Delta x$$

$$\frac{E_y\left(x+\frac{\Delta x}{2}\right) - E_y\left(x-\frac{\Delta x}{2}\right)}{\Delta x} = -\frac{\partial B_z}{\partial t}, \qquad (1.38)$$

for which in the limit $\Delta x \to 0$,

$$\frac{\partial E_y}{\partial x} = -\frac{\partial B_z}{\partial t}. \qquad (1.39)$$

The variation of the magnetic field with x can be found by applying the Maxwell-Ampere equation [Eq. (1.4)] to the green loop just in front of the sheet (Ex. 1.3), with the result

$$\frac{\partial B_z}{\partial x} = -\mu_0 \varepsilon_0 \frac{\partial E_y}{\partial t}. \qquad (1.40)$$

Eq. (1.39) and Eq. (1.40) are more quickly obtained from the differential equations for Faraday's law, Eq. (1.9), and the Maxwell-Ampere law, Eq. (1.10). For those who are already familiar with multi-variable calculus, the use of Eq. (1.9) for obtaining Eq. (1.39) is outlined in Appendix A.5.

By combining Eq. (1.39) with Eq. (1.40) while taking derivatives,

$$\frac{\partial}{\partial x}\left(\frac{\partial B_z}{\partial x}\right) = -\mu_0\varepsilon_0 \frac{\partial}{\partial x}\left(\frac{\partial E_y}{\partial t}\right) = -\mu_0\varepsilon_0 \frac{\partial}{\partial t}\left(\frac{\partial E_y}{\partial x}\right) = -\frac{\partial}{\partial t}\left(-\mu_0\varepsilon_0 \frac{\partial B_z}{\partial t}\right), \quad (1.41)$$

where we have used the fact that for an analytic function, sequential partial derivatives commute. Using just the first and last terms of Eq. (1.41) reveals a wave equation,

$$\frac{\partial^2 B_z}{\partial x^2} = \mu_0\varepsilon_0 \frac{\partial^2 B_z}{\partial t^2}. \quad (1.42)$$

If we had started by taking $\partial/\partial x$ of Eq. (1.39) and inserted Eq. (1.40), a similar wave equation for the electric field arises

$$\frac{\partial^2 E_y}{\partial x^2} = \mu_0\varepsilon_0 \frac{\partial^2 E_y}{\partial t^2}. \quad (1.43)$$

The solution for the magnetic field can be arrived at by analogy with the string solution, Eq. (1.33);

$$B_z(x,t) = B_{z0}\cos(kx - \omega t + \phi), \quad (1.44)$$

where we have taken the solution to propagate toward $+x$, consistent with energy flow away from the surface on the front side of the sheet. Eq. (1.44) is known as a plane wave, since it has plane wave fronts (i.e. flat); the phase is constant perpendicular to the direction of propagation. A phase constant ϕ is added in order to match the boundary condition on the surface. In particular, at $x=0$ this this solution must match Eq. (1.36), which requires that $\phi=0$, and $B_{z0} = \mu_0 K_0/2$; the magnetic field is then

$$B_z(x,t) = (\mu_0/2)K_0\cos(kx - \omega t). \quad (1.45)$$

This equation has an interesting interpretation. It can be written equivalently as

$$B_z(x,t) = (\mu_0/2)K_0\cos[\omega(t - x/v)], \quad (1.46)$$

where v is the speed of the wave; $v = \omega/k$. At $x = 0$, Eq. (1.46) agrees with Eq. (1.36), but at some arbitrary x it is the source current at an earlier time $t' = t - x/v$ that counts. So the time t' is termed "retarded time"; the current that generates the field at position x and time t occurred at an earlier time $t - x/v$.

13

That is because of the delay caused by the finite speed of this electromagnetic wave.

The wave solution to Eq. (1.43) for the electric field along y has the same phase as the magnetic field in Eq. (1.36) as guaranteed by Eq. (1.39),

$$E_y(x,t) = E_{y0}\cos(kx - \omega t). \tag{1.47}$$

Fig. 1.5 shows an electromagnetic wave propagating from the current sheet.

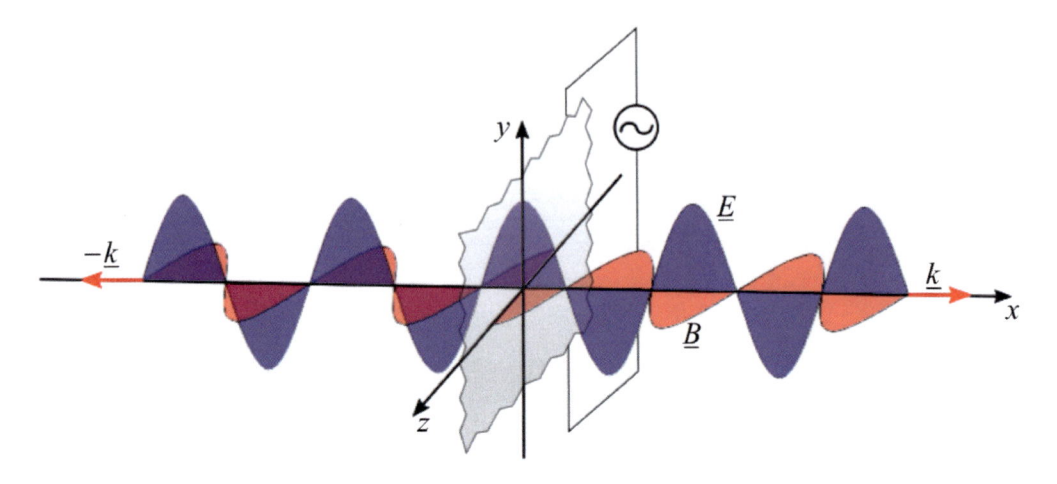

Figure 1.5 – Electromagnetic radiation propagating along the x-axis; the electric field is in the y-direction and the magnetic field is in the z-direction.

E_{y0} and B_{z0} are the amplitudes of the electric and magnetic fields. The sheet not only generates waves propagating away from its front side, but also from the backside. Whereas the speed at which a wave on a string can propagate is governed by the tension and linear density of the string [Eq. (1.26)], the generic form of the wave equation [Eq. (1.25)] when compared to Eq. (1.43) reveals the speed of electromagnetic waves is determined by μ_0 and ε_0,

$$v = \frac{1}{\sqrt{\mu_0 \varepsilon_0}}. \tag{1.48}$$

Upon substituting, $\mu_0 = 4\pi \times 10^{-7}\, T \cdot m/A$ and $\varepsilon_0 = 8.85419 \times 10^{-12}\, C^2/Nm^2$ we find the speed is, $v = 2.99792 \times 10^8\, m/s$. Maxwell realized that known terrestrial velocities paled in comparison with this and began to investigate how this number related to the physical world. To put this speed into perspective, the fastest domestic travel between NY and Los Angeles takes over 5 hours, but at a speed of $3 \times 10^8\, m/s$ that time is cut down to 13 milliseconds! The waves produced were identified as electromagnetic waves but it was not immediately obvious that they were related to light. At the time of Maxwell's discovery some progress had been made in measuring the speed of light. We will now transition to a discussion of experimental approaches for measuring the light's speed along with results.

1.5 – Measuring the Speed of Light

The first known attempt to measure the speed of light was made in the early 1600s by Galileo Galilei. His approach was to measure the speed of light using two lamps with shutters on hilltops about 1 mile apart. To start the experiment Galileo would open the shutter of his lamp on one hilltop, and when his assistant on the other hilltop saw the light from Galileo's lamp he would open the shutter on his lamp. Galileo recorded the time from which he opened his lamp shutter to the time when he saw the light from his assistant's lamp. Since he knew the distance between the two lamps he could estimate the speed at which light traveled. After repeated attempts with increasing distances between the two participants, Galileo correctly realized that light traveled too fast to be measured via this method. Without precision clocks and highspeed shutters, Galileo's limitations would be overcome by using a much-much greater distance; interplanetary.

In 1676 Ole Rømer observed that there were variations in the apparent orbital period of Jupiter's moon, Io, as measured on Earth depending on whether the Earth were moving towards Jupiter or away from Jupiter. He measured the orbital period of Io by observing the time difference between successive eclipses. Rømer hypothesized that the maximum discrepancy could be used to measure the speed of light provided one knew the difference in the distance between the nearest and farthest points in the orbit. Fig. 1.6 is Rømer's drawing, which illustrates this idea.

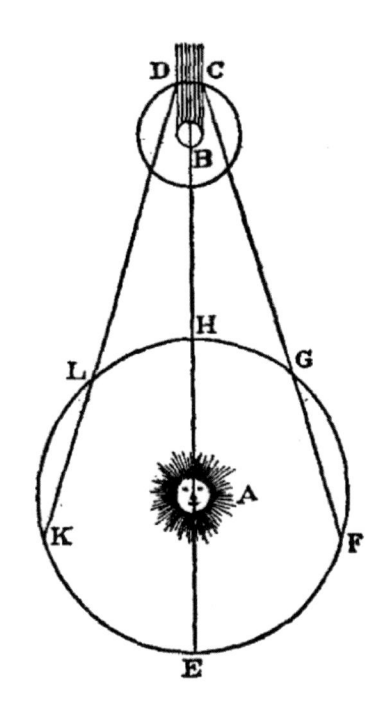

Figure 1.6 – A diagram by Rømer of the Earth's position relative to Jupiter as it orbits the Sun.

15

In Fig. 1.6, the Sun lies at A, and Jupiter at B. The circle around the Sun is the Earth's orbit and the shadow extending behind Jupiter marks the region in which moon is eclipsed. If the Earth orbits counterclockwise around the Sun then as it travels from point F to G in its orbit the time of the eclipse is shorter than as it travels from L to K. This is because as the Earth travels from F to G it is moving towards Jupiter and the distance light from Io has to travel to G is less than the distance to F. From L to K the Earth travels in the opposite direction and is closer at L than at K. By measuring the delay and the difference in distance one should be able to discover the speed of light. At the time of his discovery Rømer did not know the diameter of the Earth's orbit and was therefore unable to make an estimate of the speed of light. Using Rømer's method and other experimental techniques, scientists before Maxwell were able to measure the speed of light to be about $2.98 \times 10^8 \, m/s$. The experimental measurements available at the time allowed Maxwell to confirm that the electromagnetic waves predicted by his equations were in fact light.

Terestial measurements of the velocity of light followed in the 20th century by using accurate timers and high-speed shutters. The most notable of these was the measurement of Albert A. Michelson (Michelson, 1927).

In 1927 Albert Michelson measured the speed of light with much greater accuracy. The apparatus he used is illustrated in Fig. 1.7 and features an eight-sided mirror that is rotated with angular frequency ω.

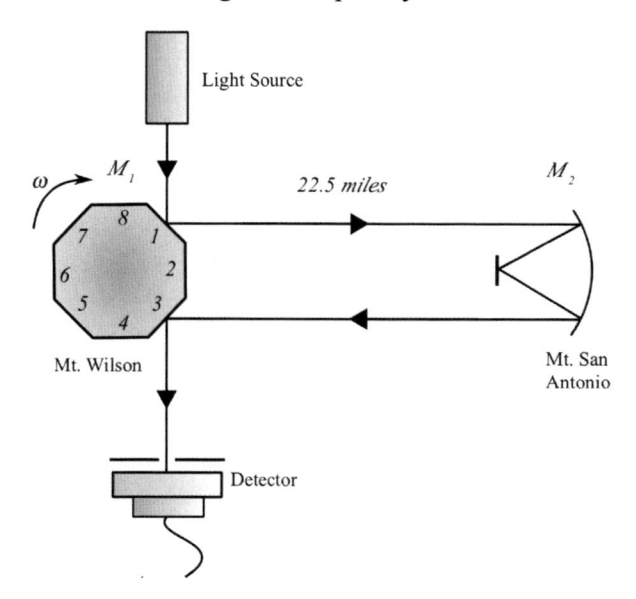

Figure 1.7 – The apparatus designed by Albert Michelson to measure the speed of light.

A light source at Mt. Wilson is reflected off side 1 of the mirror and travels 35425.15 ± 0.01 m to Mt. San Antonio where it is reflected back to Mt. Wilson. If the mirror were not rotating, i.e. $\omega = 0$, the returning light would be incident on side 3 and reflect into the detector. Of course, this would give no information about the speed. To find the speed we need to find the time it takes the light to travel from the source to the detector; we already know the distance traveled and the speed is simply the distance divided by the time. When we rotate the eight-sided mirror the light intensity on the detector changes. This is because light incident on the mirror at an angle other than 45° will not reflect directly to the detector and at some angles will not reach the detector at all. Recall from your introductory courses that the angle of incidence is equal to the angle of reflection; if you examine the geometry in Fig. 1.7 it is clear that the intensity at the detector varies with the angles at which light hits the mirrors. Now consider the mirror to be rotating with constant angular velocity. At some angular velocity the light may be incident on side 1 at 45° and by the time it returns to the mirror be incident on side 2 at 45°, this would produce the maximum intensity. The least angular frequency for this situation to occur would require the mirror to rotate an angle $\theta = 2\pi/8 \, rad$ in the time it takes for light to travel 70850.3±0.2 m. By measuring the intensity at the detector Michelson found for one of his experiments that this occurred when ω = 3.318 krad/s. The corresponding travel time is $t = \theta/\omega =$ 0.0002367 s, from which the velocity of light in air is $v_a = d/t = 299,325 \pm 4$ km/s. Michelson corrected this for the fact that light moves slightly slower in air than in vacuum due to the refractive index of air n_a being slightly larger than 1; $c = n_a v_a$.[5] This added an additional 67 km/s to the speed bringing it to 299,392 km/s. Further corrections were made by synchronizing the time measurement associated with the rotational period of the octagonal mirror to other standard oscillators. Beyond this the octogon sides were distorted causing an additional correction to be made. The final result for the velocity of light in vacuum from Michelson's paper in 1927 was 299,796 km/s (Michelson, 1927). This overestimates the speed of light by approximately one part in 100,000. Notice that

---

[5] The origin of refractive index n will be discussed more extensively in Chap. 3.

the principle is very similar to that of Galileo and was made possible by great improvements in technology; the mirror rotated at about 31,000 rpm (rotations per minute), this is about three times the rpm of the fastest modern internal combustion car engine.

The speed of light took on greater significance in the early part of the 20th century than simply the velocity of an arbitrary thing. Einstein's second postulate of special relativity required that the speed of light be invariant with respect to the motion of the source. It also predicted that no material object or information transfer could exceed this speed. Experiments over the subsequent 100+ years have not violated this postulate, or its speed limit consequence. In current times the speed of light in vacuum is recognized as one of the few universal constants. For that reason the normal symbol for velocity v, is denoted by c for the velocity of light in vacuum, with it being the first letter in the Latin word for speed, *celeritas*. In reverence to Maxwell's great accomplishment we repeat and highlight his equation,

$$c = \frac{1}{\sqrt{\mu_0 \varepsilon_0}}. \tag{1.49}$$

1.6 – Electromagnetic Waves in Three Dimensions

Traveling waves are waves that maintain their form as they move through space. These waves travel with a constant velocity, which can be measured by observing the speed of a wave crest relative to some fixed observation point. Traveling waves can appear in many different forms, as can be seen in Exercise 1.1.

A wave traveling through space moves in three dimensions, and not simply along the x-axis as depicted in Fig. 1.2. For example, we could have a wave that travels along at some angle from the x-axis, but in the *xy*-plane as seen in Fig. 1.8, and with oscillations in and out of the page (i.e. z-axis). If we choose a wave with a cosine form the electric field is given by

$$\underline{E}(\xi,t) = E_0 \hat{z} \cos(k\xi - \omega t). \tag{1.50}$$

where ξ is the coordinate along which the wave travels and underlining denotes a vector.

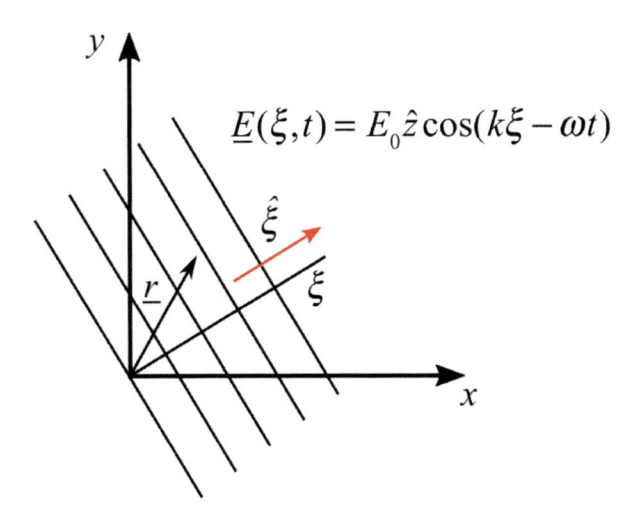

Figure 1.8 – A wave propagating along the xy-plane. The parallel lines represent the crests of the waves.

To generate a 3D description of our wave we invent a unit vector $\hat{\xi}$ in the direction of travel (i.e. $\hat{\xi}$ has no dimensions and its length is one). This is a convenient device since it will enable us to obtain an equation for a wave travelling in an arbitrary direction. Our aim is to describe the wave's displacement at an arbitrary position $\underline{r}$, see Fig. 1.8. This vector projected onto $\hat{\xi}$ gives ξ, i.e. $\xi = \hat{\xi} \cdot \underline{r}$. We can substitute this relation into the original equation to define the electric field at position $\underline{r}$, $\underline{E}(\underline{r},t) = E_0 \hat{z} \cos(k\hat{\xi} \cdot \underline{r} - \omega t)$. To neaten things up a little we define a new vector $\underline{k} = k\hat{\xi}$, the wave propagation vector, which has the same direction as $\hat{\xi}$ and whose magnitude is the spatial frequency. The solution is now

$$\underline{E}(\underline{r},t) = E_0 \hat{z} \cos(\underline{k} \cdot \underline{r} - \omega t). \qquad (1.51)$$

This result is not general because it has the full amplitude when $r = t = 0$ and that may not be the case. The way to get around this is to incorporate an adjustable phase constant ϕ into Eq. (1.51),

$$\underline{E}(\underline{r},t) = E_0 \hat{z} \cos(\underline{k} \cdot \underline{r} - \omega t + \phi). \qquad (1.52)$$

This is a good time to add some further mathematics that will make our computations easier as we more forward. You may recall Euler's famous identity $e^{i\theta} = \cos(\theta) + i\sin(\theta)$ (If you aren't familiar with this, or need some review, refer to Appendices A.1 and A.2). In what follows we will make use of Euler complex representation. Expressing Eq. (1.52) in terms of Euler's equation gives a generalized 3D wave,

$$\underline{E}(\underline{r},t) = E_0 \hat{z} e^{i(\underline{k}\cdot\underline{r} - \omega t + \phi)}. \tag{1.53}$$

Where the electric field is directed in the z-direction as indicated by the unit vector, also known as the polarization direction. The field in Eq. (1.53) contains both real and imaginary parts. Our interest is in the real part of the field, however we maintain the imaginary part for intermediate calculations. For a wave moving in a direction with 3D components for its wave vector $\underline{k}$, at 3D position $\underline{r}$, the dot product in the exponent of Eq. (1.53)

$$\underline{k}\cdot\underline{r} = (k_x\hat{x} + k_y\hat{y} + k_z\hat{z})\cdot(x\hat{x} + y\hat{y} + z\hat{z}) = k_x x + k_y y + k_z z. \tag{1.54}$$

For a polarization, $\hat{p}$, perpendicular to $\underline{k}$, the wave has an expanded form

$$\underline{E}(\underline{r},t) = E_0 \hat{p} e^{i(k_x x + k_y y + k_z z - \omega t + \phi)} \tag{1.55}$$

As an example of its use we return to our special case: A wave traveling in the x-direction with its electric field along y and its magnetic field along z, Fig. 1.9.

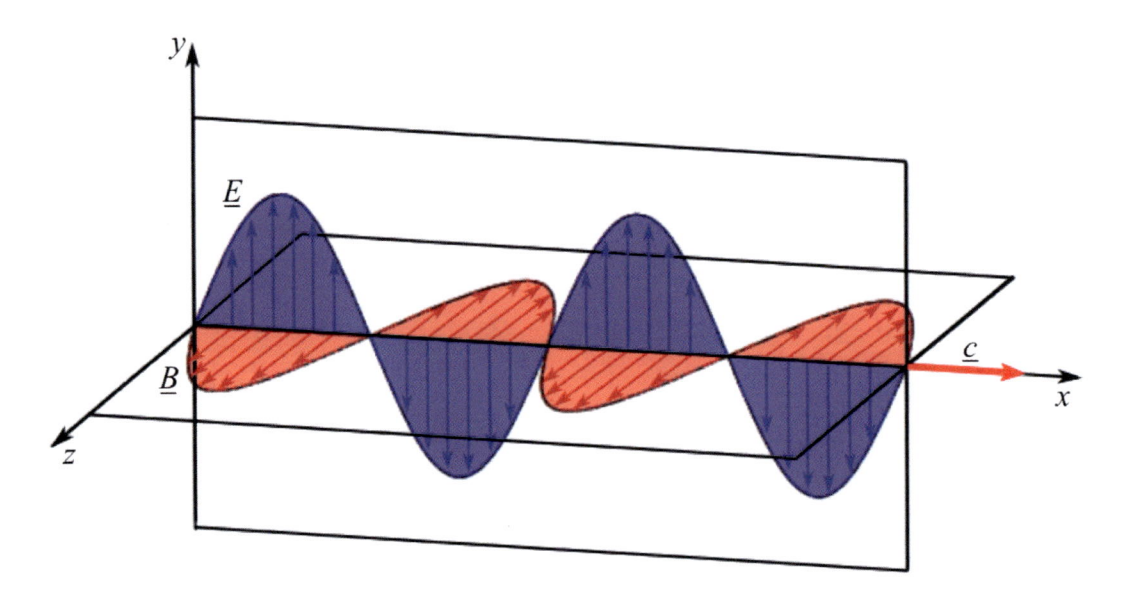

Figure 1.9 – Snapshot of an electromagnetic wave described by Eq. (1.56) traveling along the x-axis at $t = 0$ with $\phi = -\pi/2$.

In complex form these waves are

$$\underline{E}(x,t) = E_0 \hat{y} e^{i(kx - \omega t + \phi)} \text{ and} \tag{1.56}$$

$$\underline{B}(x,t) = B_0 \hat{z} e^{i(kx - \omega t + \phi)}. \tag{1.57}$$

Now let's ask what physics demands for the relationship between E_0 and B_0, while using these complex waves. Once again we return to

$$\frac{\partial E_y}{\partial x} = -\frac{\partial B_z}{\partial t},$$

which was derived from Faraday's Law [Eq. (1.9)]. By substituting the electric and magnetic waves, Eqs. (1.56) and (1.57), into this equation you should be able to show that

$$kE_0 = \omega B_0. \tag{1.58}$$

However,

$$\frac{\omega}{k} = c, \text{ and therefore } B_0 = \frac{E_0}{c}. \tag{1.59}$$

So the complex equations give the same result as Exercise 1.3, which is arrived at using real expressions.

1.7 – The Electromagnetic Spectrum

Electromagnetic waves vary over a wide range of frequencies. Fig. 1.10 shows the range as it relates to wavelength and frequency. Throughout this text all categories from Gamma rays to Radio waves are referred to as light, however each have different applications.

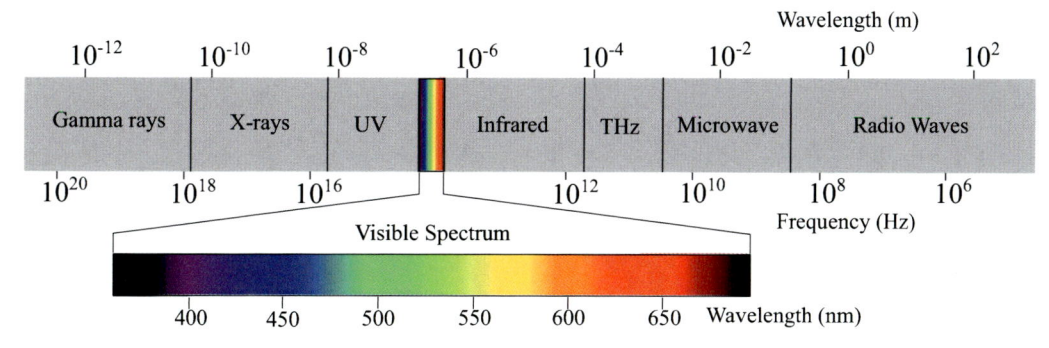

Figure 1.10 – The electromagnetic spectrum. The visible range only makes up a small portion of the total spectrum.

Beginning with shortest wavelength in Fig. 1.10 we have gamma radiation that includes all wavelengths less than 0.01 nm and frequencies higher than 30 exahertz (10^{18} Hz). Gamma radiation has extremely high energies and can be harmful to humans. High-energy radiation can increase risk of cancer by causing damage to cells and genetic material. The same property allows gamma rays to be used in surgery to kill cancer cells and prevent the spread of cancer through the body. Waves ranging from 0.01 nm to 10 nm in wavelength are called X-rays. At their associated frequency X-rays can penetrate the human body. This makes X-rays extremely useful for medical imaging. Next, lies ultraviolet (UV) radiation with wavelengths ranging from 10 nm to 390 nm. The visible spectrum ranges in wavelength from 390 nm to 700 nm, the range of electromagnetic radiation that humans can detect without assistance from ancillary machinery. Next we have infrared radiation that ranges from 700 nm to 100 μm in wavelength and from 3 to 430 THz in frequency. Infrared radiation is used in astronomy and for night vision goggles as the human body radiates primarily in the infrared range. After infrared we come to the terahertz frequency band, which ranges from 300 GHz to 3 terahertz (THz). Terahertz radiation is said to occupy the sub-millimeter band and is used in astronomy and security screening. Microwaves have shorter wavelengths than radio waves, ranging from 1 mm to ~30 cm with frequencies ranging from 1 to 300 gigahertz (GHz). Microwaves are used in communication devices and radar; they are also used in microwave ovens to heat food. Last are radio waves with wavelengths ranging from ~30 cm to hundreds of meters. These waves are typically used for communication devices such as television and radio. A wide range of applications accompanies the wide range of frequencies in the electromagnetic spectrum. The entire spectrum will be termed "light" with the narrow band perceived by humans with unaided eyes as "visible light". Scientists and engineers have found ways to produce and control radiation over a range of frequencies. In the following chapters we will describe some of these technologies, but first we need to understand more about the properties of light.

1.8 – Energy Density and Intensity of Light

In order to describe electromagnetic radiation better we are now going to discuss the energy contained within the fields and the forces exerted by light. Doing this will allow us to describe how light interacts with matter and to see why describing light in terms of frequency and wavelength is so important. To start we examine the energy stored in the electric and magnetic fields.

The energy stored in the electric field can be calculated by examining a capacitor. The charges stored on the parallel plates of the capacitor produce an electric field as seen in Fig. 1.11. We are interested in the field between the plates and would like to relate this to the energy stored in the capacitor. To begin recall that the capacitance is

$$C = \frac{Q}{V} \qquad (1.60)$$

Here Q is the charge on the plates, and V is the voltage across the plates.

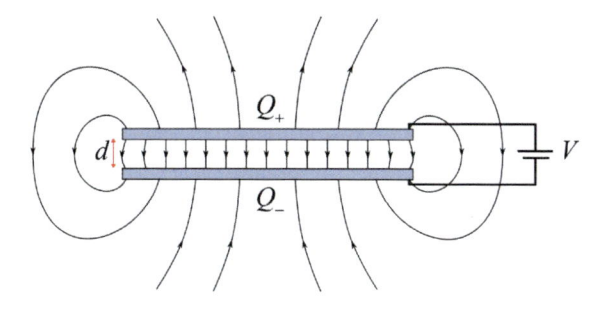

Figure 1.11 – A capacitor and the electric field it produces.

The energy stored in the capacitor is associated with plate separation d. In the process of charging, electrons are moved from the top plate to the bottom. Each increment of charge dq moved through the potential difference V requires work Vdq. As each plate takes on a net charge, V increases by q/C. The energy build up is then

$$\varepsilon = \int_0^Q \frac{q}{C} dq = \frac{Q^2}{2C}, \text{ or} \qquad (1.61)$$

$$\varepsilon = \frac{CV^2}{2}. \qquad (1.62)$$

From introductory physics we know that a uniform field E is proportional to the potential difference V and inversely proportional to the plate separation; $E = V/d$. We also know that for a parallel plate capacitor, the capacitance is proportional to the ratio of the surface area A of each plate and inversely proportional to the separation; $C = \varepsilon_0 A/d$. By substituting for V and C on the right-hand side (RHS) of Eq. (1.62), we find that

$$\varepsilon = \frac{\varepsilon_0 A}{2d}(Ed)^2 = \frac{\varepsilon_0 A E^2 d}{2}. \qquad (1.63)$$

Recognizing Ad as the volume occupied by the field we see that the energy per unit volume, or energy density, is

$$u_E = \frac{\varepsilon_0 E^2}{2}. \qquad (1.64)$$

Even though light traveling through space is not generated between metallic plates the previous equation for u_E still represents the energy density in the electric field. Light is a dynamo interconverting electric energy and magnetic energy, and for that reason there must be equal amounts of energy in each. Thus the total energy density u is just twice u_E;

$$u = u_E + u_B = \varepsilon_0 E^2. \qquad (1.65)$$

We can define the power associated with radiation by considering some volume of energy that washes over an observer in a period of time. This can be visualized using Fig. 1.12.

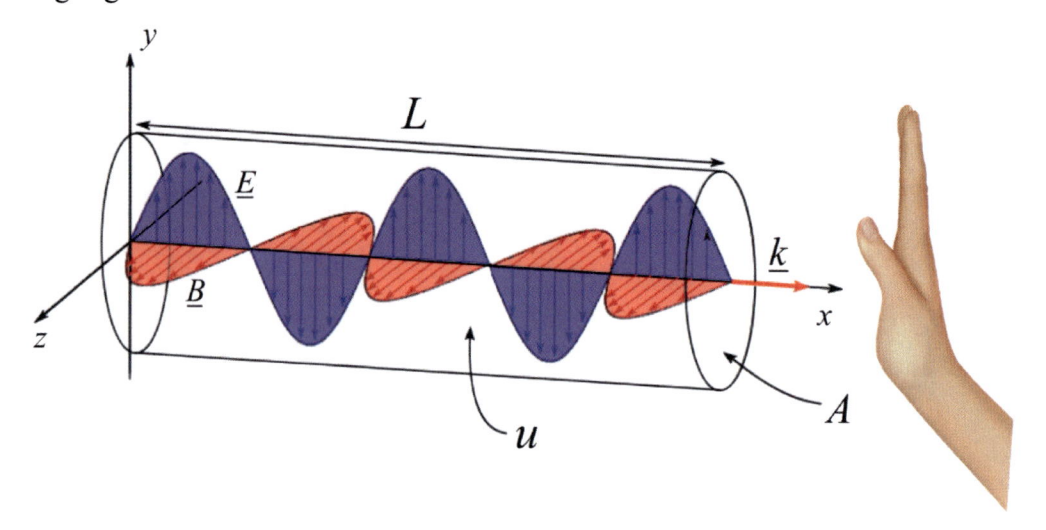

Figure 1.12 – An electromagnetic wave contained in a cylinder.

The cylinder of energy in Fig. 1.12 has length L and cross-sectional area A. The energy contained in the cylinder $\mathcal{E}$ is the energy density times the volume; $\mathcal{E} = uAL$. The electromagnetic wave is moving at the speed of light as it impacts the detector, such as the outstretched hand shown, in time $T = L/c$. Thus the total power, or energy incident on the hand per unit time, is

$$P = \frac{uAL}{L/c} = uAc. \qquad (1.66)$$

The intensity of the light is the power per unit area[6];

$$I = \frac{uAc}{A} = uc. \tag{1.67}$$

You may recall that the energy density is related to the electric field, which varies rapidly in time (Eq. (1.65)). The oscillations of the electric field are so rapid that we are not interested in measuring the intensity of the field at a given time but rather its time average, $\langle I \rangle_t$,

$$\langle I \rangle_t = \varepsilon_0 \langle \text{Re}[\underline{E}(t)] \cdot \text{Re}[\underline{E}(t)] \rangle_t c, \tag{1.68}$$

where the energy density u has been substituted from Eq. (1.65), and we have assumed that the field is in its complex form, for which the actual field is its real part (see Sec. 1.6).

The real part of the electric field varies sinusoidally (Eq. (1.56)) and the average of the square of a sinusoidal function over a whole number of cycles is just ½. Therefore the time average of the square is

$$\langle \text{Re}[\underline{E}(t)] \cdot \text{Re}[\underline{E}(t)] \rangle_t = E_0^2 \langle \cos^2(kx - \omega t) \rangle_t = \frac{E_0^2}{2}. \tag{1.69}$$

Therefore the time-averaged intensity in vacuum is

$$\langle I \rangle_t = \frac{\varepsilon_0 E_0^2 c}{2}. \tag{1.70}$$

A short-cut to calculating time-averaged intensities of complex harmonic waves is to recognize that $\langle \text{Re}[\underline{E}(t)] \cdot \text{Re}[\underline{E}(t)] \rangle_t = \text{Re}[\underline{E}^* \underline{E}]/2$, as shown in Sec. A.3, where $\underline{E}^*$ is the complex conjugate of $\underline{E}$ (i.e. reverse the sign of i). If the wave is moving through a medium where the refractive index is n_m, the right hand side of Eq. (1.70) must be multiplied by the refractive index;

$$\langle I \rangle_t = \frac{\varepsilon_0 n_m E_0^2 c}{2}. \tag{1.71}$$

The intensity of sunlight normal to the Earth's surface is about 1050 W/m². We can use this value to determine the energy incident on a solar panel. If the solar panels cover an area of 10 m² then the incident power is

---

[6] Often light intensity is referred to as the magnitude of the Poynting vector.

$$P = IA = 1050 \ W/m^2 \cdot 10 \ m^2 = 10500 \ W. \tag{1.72}$$

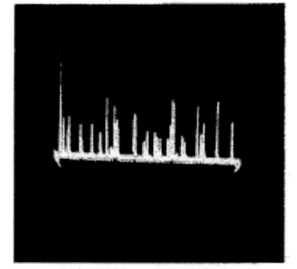

Energy is typically measured in kilowatt-hours and is given by

$$\mathcal{E} = \frac{P \cdot t}{1000}, \tag{1.73}$$

where t is the time in hours. Assuming 5 hours of direct sunlight per day the total energy incident on our solar panel system is

$$\mathcal{E} = 52.5 \ kWh. \tag{1.74}$$

Unfortunately, solar panels are only about 25% efficient, so one can expect 13 kWh per day. A Tesla electric vehicle requires about 1/4 kWh to go one mile, so the energy produced in one day can allow the car to travel 52 miles.

Surprisingly the energy in a beam of light can be thought of as carried by particles (Einstein, 1905) known as photons. Evidence for photons is easy to demonstrate in modern times for weak light sources. Fig. 1.13 illustrates the detected signal from weak light by a very sensitive photomultiplier. As you can see, spikes are recorded, which arise from particles hitting the photomultiplier's cathode. This is an opportune time to discuss photons for the purpose of eventually talking about the momentum of light.

Figure 1.13 – Article showing the time trace of current from a photo-detector in the presence of very weak light. The spikes may be interpreted as individual photons (Arnold, 1977).

1.9 – Photons, Momentum of Light, and Forces Exerted by Light

Photons were invented by Einstein to explain the photoelectric effect. His theory of relativity provides a means for describing the energy $\mathcal{E}$ of any particle in the universe,

$$\mathcal{E}^2 = p^2c^2 + m_0^2 c^4, \qquad (1.75)$$

where p is momentum and m_0 is the rest mass. Since photons move at the speed of light they have no rest mass; $m_{0,photon} = 0$. However, Eq. (1.75) reveals that photons have momentum

$$p = \pm \frac{\mathcal{E}}{c}, \qquad (1.76)$$

for a photon traveling to the right or left. Einstein recognized the energy of a photon is proportional to a constant times the frequency of the light wave

$$\mathcal{E} = \hbar\omega, \qquad (1.77)$$

where $\hbar$ is Planck's constant h divided by 2π and has the value

$$\hbar = 1.056 \times 10^{-34} \, J \cdot s. \qquad (1.78)$$

Combining Eq. (1.76) with Eq. (1.77) we find the momentum of a single photon is

$$p = \frac{\mathcal{E}}{c} = \frac{\hbar\omega}{c} = \frac{h}{\lambda}. \qquad (1.79)$$

Now we are able to express the energy and momentum of single photons in terms of frequency and wavelength. Eq. (1.79) has important consequences. Let's begin by considering the transfer of momentum from photons to a reflecting surface, Fig. 1.14.

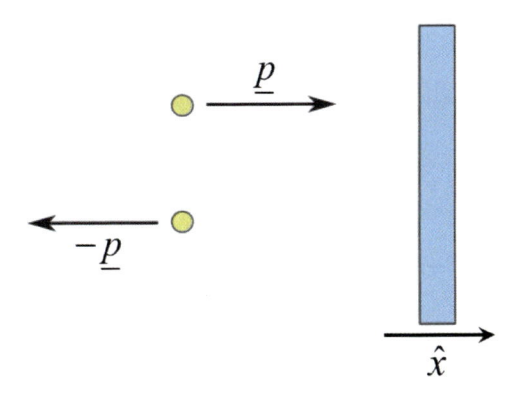

Figure 1.14 – The momenta of a photon before and after it has been reflected off a surface.

After the reflection shown in Fig. 1.14 the change in the photon's momentum is $-2\underline{p}$. We would like to know the force exerted by these collisions. This comes from Newton's second law; force is the time-rate change of momentum. The time-average force is the rate of photon collisions times the change in momentum per collision. If a large number of photons strike the surface with rate R_p, the time-averaged force of the surface on the photons $\langle \underline{F} \rangle_t$ is $\langle \underline{F} \rangle_t = R_p(-2\underline{p})$. This force is in the negative $\hat{x}$ direction. The force on the mirror is it's reaction; it is towards the right and given by Newton's third law; $\langle \underline{F} \rangle = R_p(+2\underline{p})$. Using Einstein's equation for energy, Eq. (1.76), we can write this as

$$\langle \underline{F} \rangle = 2R_p \left(\frac{\mathcal{E}}{c} \right) \hat{x} \qquad (1.80)$$

Examining Eq. (1.80) closely we see that $R_p \mathcal{E}$ is the incident power, i.e. the rate of energy being deposited;

$$P = R_p \mathcal{E}. \qquad (1.81)$$

The force exerted on the mirror is then

$$\langle \underline{F} \rangle = \frac{2P}{c} \hat{x}. \qquad (1.82)$$

The factor of 2 is missing is the case of total absorption since there is no recoil momentum in this case.

We have developed a mechanism for dealing with light as a photon and as a wave. In this text we will use both representations and will often attempt to explain phenomena in a photon, or particle point view, as well as that of a wave. Some things, such as diffraction are better treated with light represented as a wave, while circular polarization is easier to explain for a photon.

We can see how momentum of a photon or beam of light can be transferred to spherical glass bead by analyzing how a spherical dielectric interacts with the radiation. A glass sphere such as the one shown in Fig. 1.15 will bend (refract) a beam of light that travels through it. It is clear that the direction in which the light is propagating is changed by its interaction with the sphere. The momentum vector $\underline{p}_{in}$ shown in Fig. 1.15 has also changed by amount $\Delta \underline{p}$. We know from mechanics that momentum is conserved so the photon must have transferred momentum $-\Delta \underline{p}$ to the sphere.

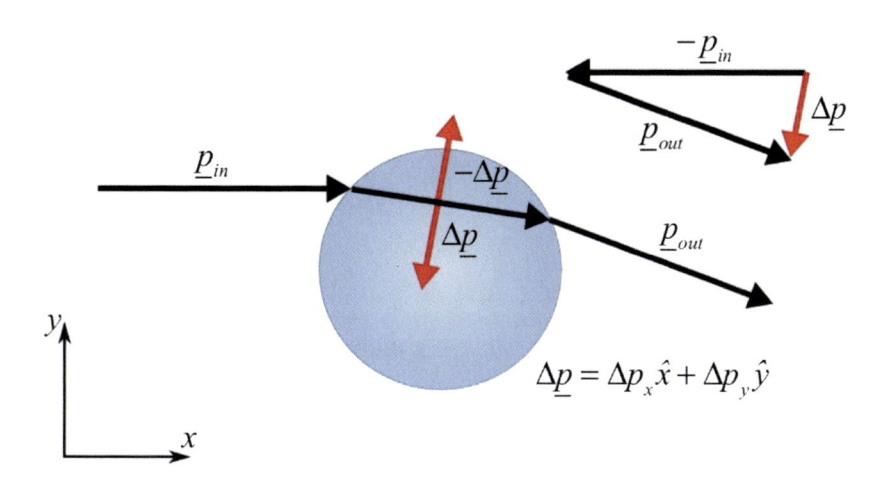

Figure 1.15 – The bending of light by a spherical lens. The incident photons transfer momentum to the sphere.

The transfer of momentum means that the sphere is exerting a force on the photon, and by Newton's Third Law that the photon is exerting an opposite force on the sphere. Newton's Second Law governs the force exerted by a single photon;

$$\underline{F}_p = -\frac{\Delta \underline{p}}{\Delta t}. \tag{1.83}$$

The time-averaged force on the sphere from a beam of photons is

$$\langle \underline{F} \rangle_{tot} = -R_p \Delta \underline{p}. \tag{1.84}$$

You may have noticed that the dielectric in Fig. 1.15 is spherically symmetric so if it were placed in a uniform beam of light in the x-direction, the forces in the y-direction would all cancel. Still the light beam would impart some force in the forward x-direction on the particle. In the next section we will discuss optical tweezers, which are an application of the physics we have just described.

1.10 – Optical Levitation and Optical Tweezers

Up to this point we have only discussed fundamental aspects of light. Now we would like to describe an application of what we have learned, namely, optical tweezers. To warm up, we will analyze the effect a beam of many photons has on a dielectric particle.

The forces exerted by photons on an object, such as the dielectric sphere in Fig. 1.15, can be used to manipulate it. These forces can be used to levitate a dielectric placed in a light beam. The particle floats in space, or levitates, when the upward force on the dielectric from the beam of light balances the force of gravity on the particle (Ashkin, 1971). The light beam used in optical levitation has a Gaussian intensity profile, as shown in Fig. 1.16. This figure leaves out the reflected rays that aid in levitation, and only includes refracted rays that help to explain the lateral stability of the levitated sphere.

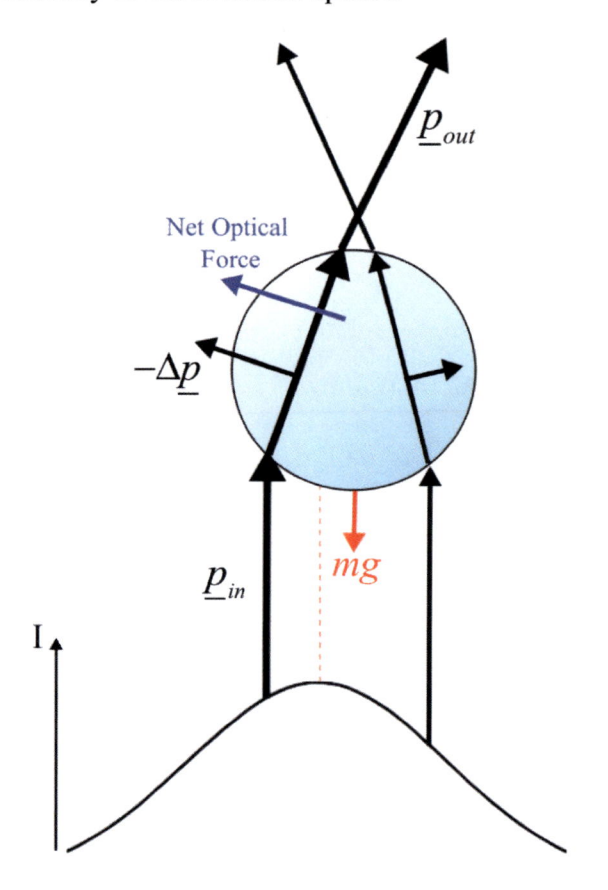

Figure 1.16 – Optical levitation of dielectric microparticle.

The Gaussian intensity profile indicates that the number of photons is greatest at the center of the beam and falls off towards the edges. Thus the rate R_p at which the photons interact with the sphere is greatest at the center of the beam. The force of the light on the sphere is proportional to the rate of photons moving through it.

Thus, when the dielectric particle is off from the beam's center, as shown in Fig. 1.16 it is pushed back towards the beam's center.

Fig. 1.17 shows an image of an optically levitated glass micro-sphere. It looks like a star as it scatters light from the levitating laser. So long as the power of the laser is constant the sphere remains in place! Actually, even electrostatic fields can generate forces on a neutral dielectric particle.

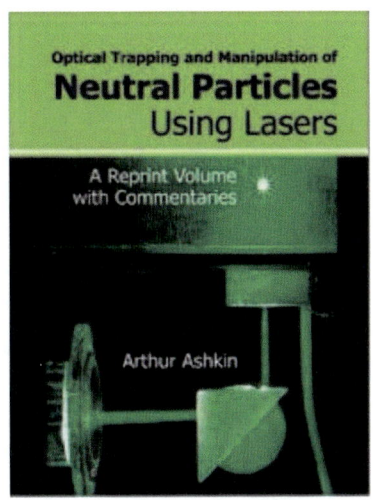

Figure 1.17 – Cover of a book on optical trapping by Arthur Ashkin.

Imagine we place a particle with no net charge in an electric field set-up by a spherical electrode wired to a circular plate with a battery, as in Fig. 1.18. How do we expect the particle (blue) to move through the field after being dropped? At first glance we may suspect since the particle has no net charge, it will be unaffected by the field and will fall because of the force of gravity. Of course physics is an experimental science so the best way to know what will happen is by conducting an experiment.

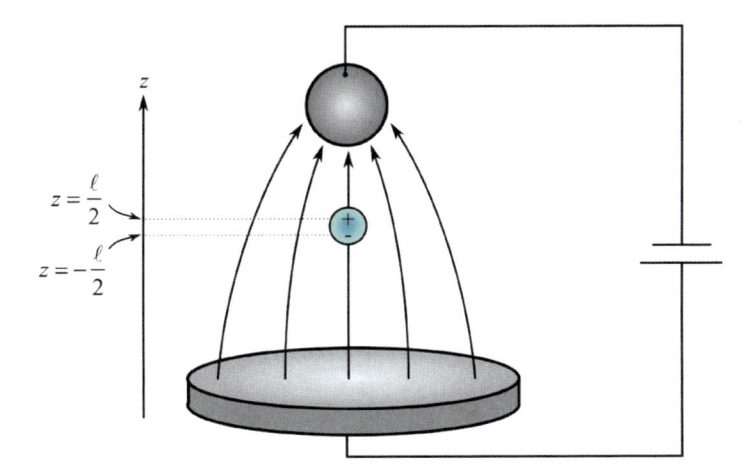

Figure 1.18 – A particle with no net charge is introduced to an electric field.

Now we introduce the particle into a strong field and observe that it moves up and hits the spherical conductor. At this point it is clear that even though the particle is neutral it is affected by the electric field. What happens if we reverse the direction of the field by reversing the battery, so that the bottom plate is now negatively charged? When we introduce the particle into this reversed system we observe that it still moves up and hits the spherical conductor! Now we start to think about what could be happening to produce this strange behavior. We know that the magnitude of the electric field in the z-direction $|E_z(z)|$ increases with z, and the field will polarize our dielectric particle as in Fig. 1.19 (shown for an upward pointing field).

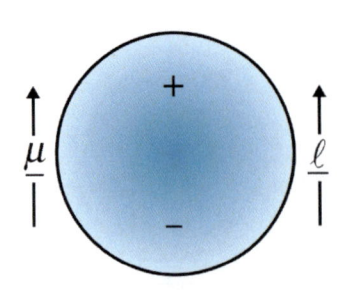

Figure 1.19 – An induced dipole.

The dipole has a dipole moment $\underline{\mu}$ and polarizability α. Polarizability describes how strong a dipole is induced by an electric field. The dipole moment, polarizability and charge separation are related as

$$\underline{\mu} = \alpha \underline{E} = q\underline{\ell}. \tag{1.85}$$

The dipole is of length ℓ and the electric field produces a force on each of the charges. The total force on the dipole is

$$F_z = |q|[E_z]_{z+\ell/2} - |q|[E_z]_{z-\ell/2}. \tag{1.86}$$

It is clear that the separation of the charges is what produces a net electric force on the particle. In order to simplify or equation we can approximate the upper field in terms of the lower through a Taylor expansion

$$[E_z]_{z+\ell/2} = [E_z]_{z-\ell/2} + \frac{\partial E_z}{\partial z}\ell. \tag{1.87}$$

Now the total force is

$$F_z = |q|\left[[E_z]_{z-\ell/2} + \frac{\partial E_z}{\partial z}\ell - [E_z]_{z-\ell/2}\right] = |q|\frac{\partial E_z}{\partial z}\ell. \tag{1.88}$$

Referring to Eq. (1.85) we see this can also be written as

$$F_z = \mu\frac{\partial E_z}{\partial z} = \alpha E_z\frac{\partial E_z}{\partial z} = \frac{\alpha}{2}\frac{\partial}{\partial z}E_z^2. \tag{1.89}$$

Eq. (1.89) makes it clear that the total force on the induced dipole is proportional to the gradient of the square of the electric field rather than the field itself. This means that the force is independent of the direction of the field, which matches our experimental observation. Thus far we have only treated the field in one-dimension. We can generalize the result to three-dimensions by replacing the partial derivative with respect to z by a sum of three vector components

$$\underline{F}_g = \frac{\alpha}{2}\left[\hat{x}\frac{\partial E^2}{\partial x} + \hat{y}\frac{\partial E^2}{\partial y} + \hat{z}\frac{\partial E^2}{\partial z}\right]. \quad (1.90)$$

This can be written more compactly using the *Del* operator [Eq.1.6]. In this respect the force in Eq.1.90 is called the "gradient force",

$$\underline{F}_g = \frac{\alpha}{2}\nabla E^2. \quad (1.91)$$

A field that is oscillating in time can be represented as $E = E_0 \sin(\omega t)$, for which the time-averaged force $\langle \underline{F}_g \rangle_t$ is

$$\langle \underline{F}_g \rangle_t = \frac{\alpha}{4}\nabla E_0^2, \quad (1.92)$$

with the additional factor of ½ arising from taking the time average of $\sin^2(\omega t)$; $\langle \sin^2(\omega t) \rangle_t = 1/2$. The gradient force in Eq. (1.92) drives the particle in the direction of larger E_0^2.[7]

The polarizability of a particle is contributed to by atoms in its interior that have field-induced polarization (see Chap.3), and is therefore proportional to the volume of the particle V_p. The field-induced polarization of an insulating material is characterized by the material's dielectric constant $\varepsilon/\varepsilon_0$. Materials that are composed of atoms that are more easily polarized have a greater dielectric constant. Here we provide an equation for the polarizability of a spherical

---

[7] That Eq. (1.92) would work at the frequency of visible light may be puzzling to some since the time varying magnetic field produces an additional electric field. Yet Eq. (1.92) does work for a Rayleigh particle in visible light, and you should wonder why!

dielectric particle having material with dielectric constant $\varepsilon_p/\varepsilon_0$, and embedded in a medium of dielectric constant $\varepsilon_m/\varepsilon_0$,

$$\alpha = 3\varepsilon_m V_p \left(\frac{\varepsilon_p - \varepsilon_m}{\varepsilon_p + 2\varepsilon_m} \right). \tag{1.93}$$

Ideally one would like to produce a trap for holding a biological object in a medium, and away from any surface. But as hard as we try a particle having a positive polarizability will always move to one of the electrodes and denature. That turns out to be fundamental; an electrostatic field cannot produce a local 3D maximum of E^2 in free space. The dynamical field of light however solves that problem.

When a lens focuses light it gets very intense at the focal point and weaker in all directions away from this point. The use of the term focal point is a misnomer, since it implies that all of the light passes through a point. That turns out to be impossible due to diffraction, a subject that will be covered in the next chapter. The correct way to view the focal region is to assign it a size. Nonetheless, laser light intensity in its basic mode falls off in all directions away from the center of the focal region. A reasonable picture of this region is shown in Fig. 1.20.

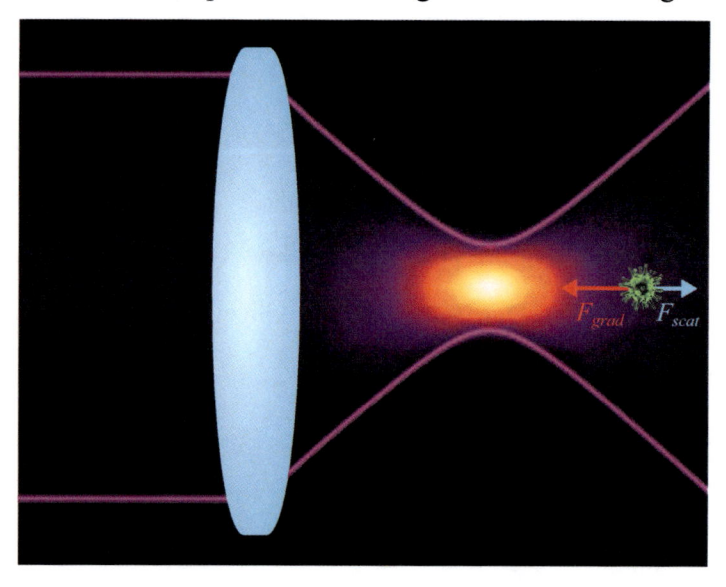

Fig. 1.20 – Depicts a virion under gradient and scattering forces just ouside the focal region. For a heavily focused laser beam the gradient force will dominate.

A dielectric particle with size much smaller than the wavelength when introduced near a focal region of a strongly focused laser beam moves toward its center, the point of highest intensity, the point at which the field squared is maximum, just as in the electrostatic case; the gradient force in Eq. (1.92) is once again working. Although an electrostatic field cannot produce a 3D extremum of an electric field in space, focused light can and the deterministic force on a dielectric particle much smaller than the wavelength of light still follows Eq. (1.92) with one exception. There is also another force along the direction of the light associated with photon scattering, $\underline{F}_s$, however, for our ultrasmall dielectric particle with radius $a \ll \lambda$, the so-called "gradient force" described by Eq. (1.92) will dominate for a strongly focused laser. The major reason for this arises from the size scaling of each of the forces. Whereas the gradient force is proportional to the volume of the particle, the scattering force is proportional to the volume squared, and therefore falls much faster as the size is reduced. Unlike the electrostatic gradient force, the optical gradient force produced by focused light can be associated with intensity by combining Eq. (1.92) with Eq. (1.71),

$$\left\langle \underline{F}_g \right\rangle_t = \frac{\alpha}{2 n_m \varepsilon_0 c} \nabla \left\langle I \right\rangle_t . \qquad (1.94)$$

The dominance of this gradient force produces a so-called "optical tweezer"; a means for moving about a very small dielectric particle without the need for destructive physical contact. This invention is attributed to Arther Ashkin and co-workers (Ashkin, 1986). A virus particle such as HIV or Influenza A is an example of such a dielectric particle (Pang, 2014). Each is about 50 nm in radius, so that light from a neodynium-doped yttrium aluminum garnet (Nd:YAG) laser with wavelength of 1064 nm is more than 20 times the radius. For future reference, each of the viruses has a volumetric polarizability near $\alpha/\varepsilon_0 = 4 \times 10^{-22} \text{m}^3$ (Arnold, 2008). Beyond optical forces, ultra-small particle trapping must recognize the importance of the thermal energy, $k_B T$, where k_B is Boltzmann's constant, and T is the absolute temperature.

To understand the role of thermal energy we must recast the gradient force in terms of potential energy. This is easy to do, since force expressed by Eq. (1.94) or Eq. (1.92) is conservative, and can be written as the negative gradient of potential energy; $\underline{F}_g = -\nabla U_g$. Comparing this equation to our equation for the gradient force (Eq. (1.94)), we see that the gradient potential is given by

$$U_g = -\frac{\alpha}{2} \frac{\langle I \rangle_t}{n_m \varepsilon_0 c}, \qquad (1.95)$$

where $\langle I \rangle_t$ is the time averaged local intensity. So the trapping potential is more negative (i.e. "deeper") where the intensity is greater. Now let's consider the profile of a focused laser beam in Fig. 1.20. A modest intensity before focusing grows to a much larger intensity at the position of the focus. The beam we will consider is the lowest order cylindrical mode and has a Gaussian intensity profile, which means that at all positions x from the focal spot (along the beam axis), the intensity perpendicular to the axis (r coordinate) follows a Gaussian function,

$$\langle I(x,r) \rangle_t = \frac{2P_{in}}{\pi w^2(x)} \exp\left(-\frac{2r^2}{w^2(x)}\right), \qquad (1.96)$$

where P_{in} is the laser power, $w(x)$ is the radius at which the intensity falls to $1/e^2$ of its axial value, and is given by

$$w^2(x) = w^2(0)\left[1 + \left(\frac{\lambda x}{\pi w^2(0)}\right)^2\right], \qquad (1.97)$$

where $w(0) \simeq \frac{\lambda}{\pi(NA)}$, and NA is the so-called numerical aperture of the focusing lens [see Eq. (10.14) and Fig. 10.10 for the definition of NA]. This is the equivalent of generating a 3D potential trap, as shown by the plot of U vs x and r. Near focus Eq. (1.97) reveals that the potential has a parabolic shape in x as well as r, as is true for all functions with such a symmetric minimum. Such a potential is associated with a linear spring cage; if the particle is moved off from the focal point it is effectively pulled back by "optical-springs". The potential in the x, r plane for an HIV virion caught in a trap by a laser beam with wavelength $\lambda_0 = 1.064 \mu m$, $P_{in} = 30 mW$, and focused by a lens with NA = 1 is shown in Fig. 1.21(b).

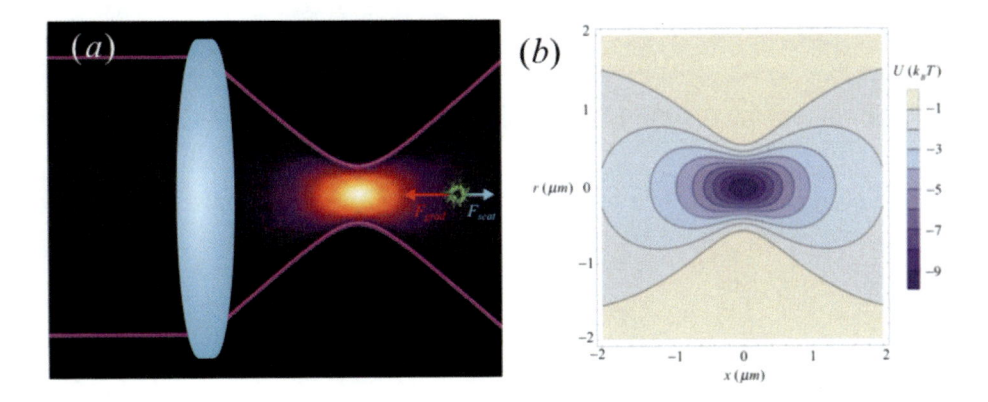

Figure 1.21 – *(a) Laser beam with wavelength* $\lambda_0 = 1.064 \mu m$ *and* $P_{in} = 30 mW$ *focused by a lens with NA=1. (b) Gradient potential energy for a HIV virion in water and within the laser beam in (a). The volumetric polarizability of HIV in water* $\alpha/\varepsilon_0 = 4 \times 10^{-22} m^3$.

For the parameters given in the figure caption, the potential at the center of the focal region is $U_g = -4.2 \times 10^{-20} J$. At room temperature a trap is produced that is 10 times deeper than the thermal energy; the virus will be trapped over an extended period.

1.11 – Chapter One Exercises

Exercise 1.1 – Determine which of the following describe traveling waves. If the equation describes a traveling wave, determine the speed and direction of motion of the wave.

(a) $\psi(z,t) = e^{-(\alpha^2 z^2 + \beta^2 t^2 + 2\alpha\beta t z)}$

(b) $\psi(x,t) = A\sin(\alpha x - \beta t^2)$

(c) $\psi(x,t) = A\cos\left[2\pi(\alpha x + \beta t)^2\right]$

(d) $\psi(y,t) = A\sin^2\left[2\pi(t-y)\right]$

Exercise 1.2 – For a plane electromagnetic wave in vacuum $E_y(x,t) = E_{y0}\cos(kx - \omega t)$, and $B_z(x,t) = B_{z0}\cos(kx - \omega t)$. By using the equation we derived from Faraday's law, $\partial E_y/\partial x = -\partial B_z/\partial t$, show that the electric field and magnetic fields are related by

$$E_y = (\omega/k)B_z.$$

Since ω/k for an electromagnetic wave in vacuum is the speed of light c, the relationship between the electric and magnetic can simply be written as $E_y = cB_z$.

Exercise 1.3 – Evaluate the integral in Eq. (1.4), $\oint_\ell \underline{B} \cdot d\underline{s} = \mu_0 \varepsilon_0 \dfrac{d}{dt} \int \underline{E} \cdot \hat{n} dA$, the Maxwell-Ampere equation in vacuum, to derive Eq. (1.40),

$$\frac{\partial B_z}{\partial x} = -\mu_0 \varepsilon_0 \frac{\partial E_y}{\partial t}.$$

Exercise 1.4 – An electromagnetic wave is specified (in SI units) by the following equation

$$\underline{E}(x,y,t) = \left(2\hat{x} + \sqrt{5}\hat{y}\right)(10V/m)\exp\left\{i\left[\frac{2}{3}\left(\sqrt{5}x - 2y\right)\times 10^8 - 6\times 10^{16} t\right]\right\}$$

Find the following:
  (a) The direction along which the electric field oscillates.
  (b) The scalar value of the amplitude of the electric field.
  (c) The direction of the propagation of the wave.
  (d) The propagation number and wavelength.
  (e) The frequency and angular frequency.
  (f) The speed of the wave.

Exercise 1.5 – Show that the result derived in Exercise 1.2, namely,

$$E_y = cB_z,$$

for a plane wave of light with $\underline{E}$ polarized along y and traveling in the x direction, can be derived from the more general equation

$$\underline{k} \times \underline{E} = \omega \underline{B}.$$

This equation essentially says that the magnetic field is perpendicular to the plane containing the wavevector and the electric field, and gives its magnitude in relation to both.

Exercise 1.6 – Show that the energy density of the magnetic field in an inductor is

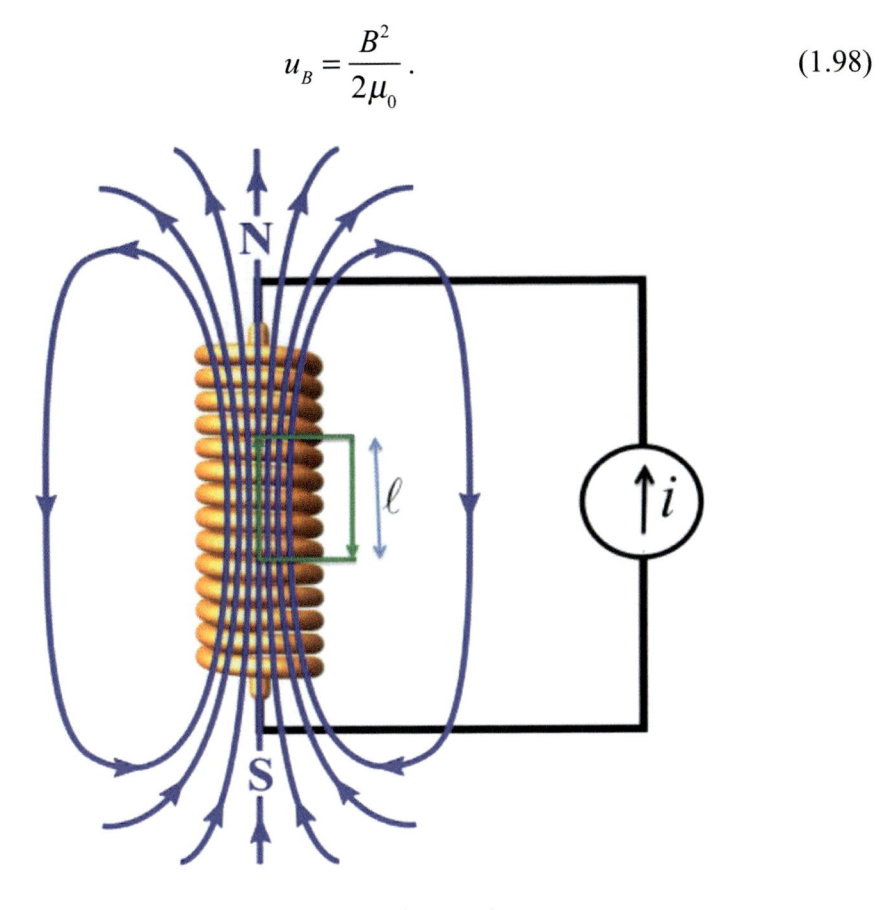

$$u_B = \frac{B^2}{2\mu_0}.\qquad(1.98)$$

Figure 1.22 – An inductor and the magnetic field it produces.

Hint: Start by evaluating the magnetic field at the center of the solenoid using the Maxwell-Ampere Equation (Eq. (1.10)), using a path integral around the green loop. Beyond this imagine slowly "charging" the solenoid up from zero magnetic field to B.

Exercise 1.7 – A linearly polarized electromagnetic harmonic plane wave with a wavelength of 600 nm and amplitude of 10 V/m is propagating in vacuum along a line in the xy-plane at 45° to the x-axis with its electric field polarized along the z-axis.

(a) Write an expression describing the wave assuming that the x and y components of $\underline{k}$ are positive.

(b) Calculate the amplitude of the magnetic field and describe its polarization.

(c) Find the time-averaged intensity of the wave.

(d) If the beam is directed at a mirror 2 cm² in area, as shown below, what is the direction and force (in Newtons) on the mirror due to radiation pressure.

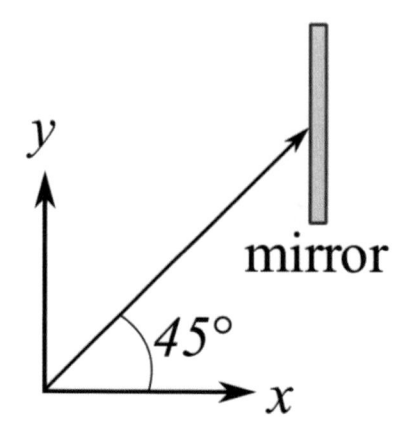

Exercise 1.8 – A laser beam of wavelength 532.0 nm and with a power of 5 mW is focused by a lens from a beam diameter of 1 cm to a beam diameter of ~1 μm over a focal length of 15 mm.

   a. What is the approximate electric field at the center of the beam just before being focused?
   b. What is the approximate electric field of the beam at the focal point?
   c. A mirror at the focal point reflects the beam directly back. What is the force of radiation pressure on the mirror? Give your answer in pico-Newtons.

* The beam diameter D is defined so that the intensity at the center is the power P divided by $\pi D^2/4$. Also, $\dfrac{1}{4\pi\varepsilon_0} = 9\times 10^9$ (MKS).

Exercise 1.9 – For the optical tweezers trap that captures a non-absorbing Rayleigh particle in Fig. 1.21 it can be shown that the ratio of the gradient force to the scattering force along the beam axis is

$$\frac{\underline{F}_g \cdot \hat{x}}{\underline{F}_s \cdot \hat{x}} = -\left(\frac{3}{32 n_m^2 \pi^3}\right)\left(\frac{\lambda_0^3}{\alpha/\varepsilon_0}\right)\frac{\pi^2 (NA)^4 (x/\lambda_0)}{\left[1+\pi^2 (NA)^4 (x/\lambda_0)^2\right]}. \qquad (1.99)$$

Find the displacement x_b for which the forces are balanced (i.e. $\underline{F}_g \cdot \hat{x} = -\underline{F}_s \cdot \hat{x}$) given the physical parameters in the caption of the figure.

Exercise 1.10 – What is the gradient potential depth relative to the thermal energy at the position of the force balance in Ex. 1.9? The thermal energy is found by multiplying Boltzmann's constant by the temperature of the medium, $E = k_B T$. Note that the power utilized in Fig. 1.21 is 30mW.

Experiment #1 - Photometry: measuring your laser's optical power

This is probably a good time to familiarize your self with your Pocket Optics kit (PO-1). Contained within are many components including a Photometer and a 650nm red laser pointer. The Photometer uses a silicon photodiode (PD) to sense light by generating a photocurrent in proportion to the absorbed power. Depending on the Photometer power range dip switch (SW1) selected, four ranges are available: 1µW, 10µW, 100µW and 1mW. Flip ON the 1mW dip switch and keep all others off for this experiment. To measure the laser beam power you will have to focus the laser so that the entire beam is within the PD's 2.65mm× 2.65mm sense area. To focus the laser beam width to a smaller size than the PD sense area requires rotating the lens focusing ring of the laser pointer .

    The goal of this experiment is to measure the power of your laser pointer. The figure below shows my experimental setup. As you can see, the laser power is focused within the photodiode (PD) sense area. Note that the luminous digital Light Sensor Display (LSD) on the Baseboard reads 3.68. To get power you multiply the LSD reading by the dip switch selected 1mW Power Range, which gives 3.68*1mW = 3.68mW. This value is correct for a wavelength near 650nm.[8]

---

[8] At other wavelengths the LSD needs to be multiplied by a Wavelength Correction Factor (WCF) to get the corresponding laser power. For example if the wavelength is 600nm for which the WCF is 1.15, a reading of 3.68 would mean a power of $3.68 \times 1.15 \times 1\,mW = 4.23\,mW$. Wavelength Correction Factors are tabulated on the back of the Photometry Board for various wavelengths.

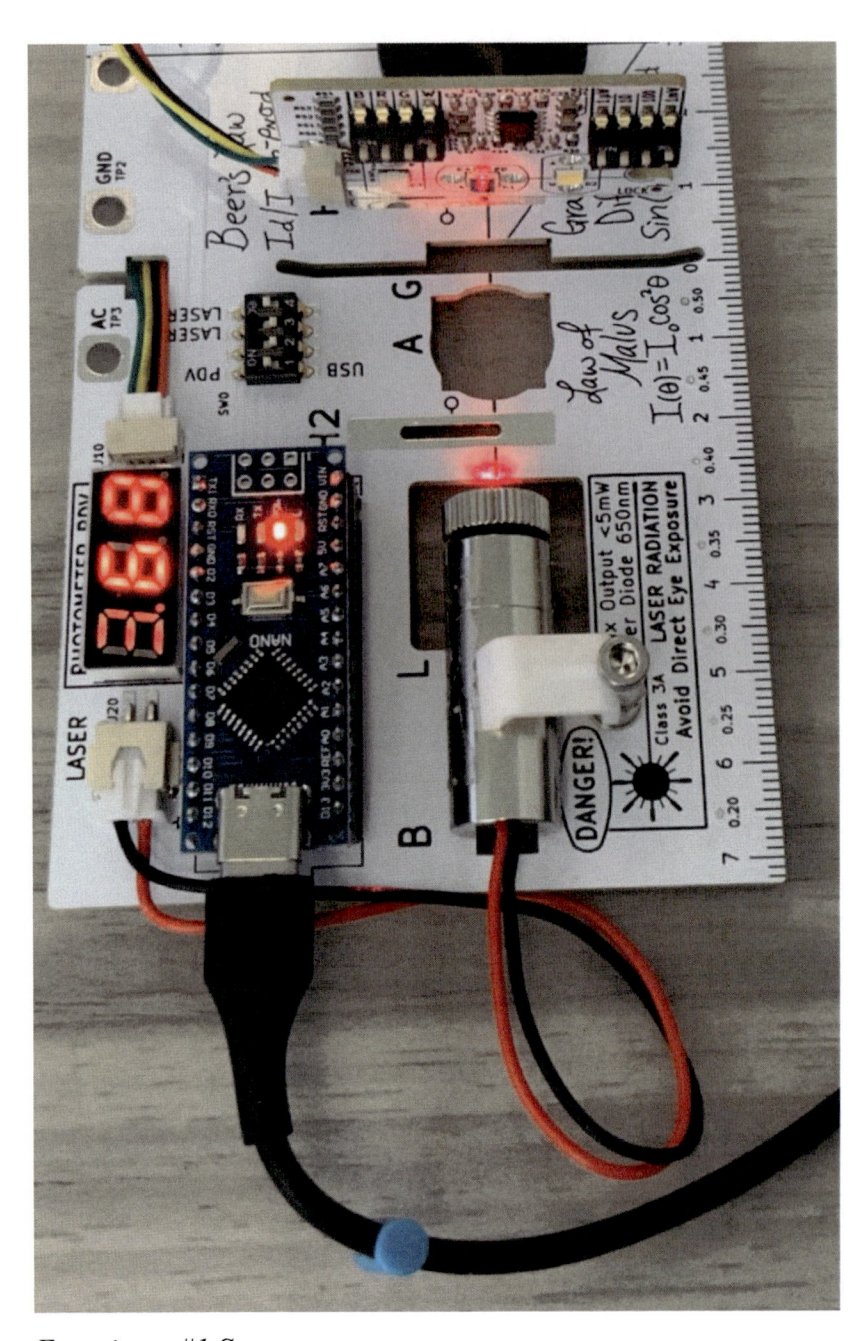

Experiment#1 Setup

Chapter Two – Diffraction

2.1 – Superposition of Waves

Now that we have developed a language with which to describe electromagnetic radiation we are going to delve into different applications. In this chapter we will discuss diffraction. Diffraction allows us to understand the structure of an object by measuring the intensity of light that has scattered off of it. The theory we will develop relies on interference patterns created by overlapping scattered waves. To start we will assume we are working with coherent light sources of a single wavelength. This assumption will make it much easier to describe interference patterns. In practice coherent sources are used for most of the experiments we will describe in this text.

Before we begin examining diffraction patterns we need to develop a powerful tool, namely the principle of superposition. The principle of superposition is a statement that the total displacement of overlapping waves is the sum of the their individual displacements as shown in Fig. 2.1.

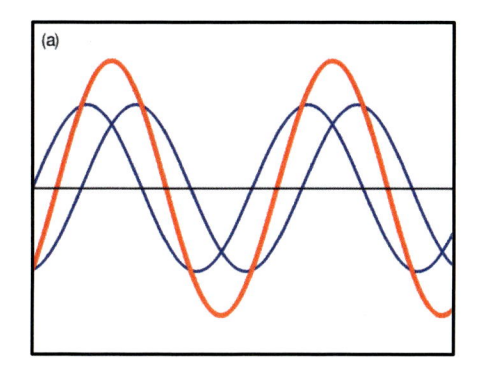

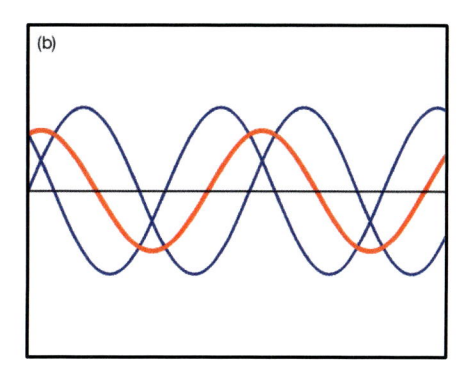

Figure 2.1 – The red waves are the sum, or superposition, of the two blue waves.

Fig. 2.1 illustrates two cases for interference in which the waves differ by an arbitrary phase shift. To find the displacements of the resulting wave (red) we merely added the displacements of the two blue waves at the corresponding position. The reason we are able to do this is because the wave equation is linear. A linear equation has two important properties. First, the sum of any two solutions is also a solution. Second, any solution times a constant is also a solution. To show that the superposition of two waves is a solution to the wave equation we first construct the wave equation separately for each of the waves that are interfering. We will call these waves ψ_1 and ψ_2;

$$\frac{\partial^2 \psi_1}{\partial x^2} - \frac{1}{c^2}\frac{\partial^2 \psi_1}{\partial t^2} = 0 \text{ and} \tag{2.1}$$

$$\frac{\partial^2 \psi_2}{\partial x^2} - \frac{1}{c^2}\frac{\partial^2 \psi_2}{\partial t^2} = 0. \tag{2.2}$$

We can add these equations together,

$$\frac{\partial^2 \psi_1}{\partial x^2} + \frac{\partial^2 \psi_2}{\partial x^2} - \frac{1}{c^2}\frac{\partial^2 \psi_1}{\partial t^2} - \frac{1}{c^2}\frac{\partial^2 \psi_2}{\partial t^2} = 0 \tag{2.3}$$

Since the 2nd derivatives are linear operators we know that the sum of the derivatives of two functions is simply the derivative of the sum of the same two functions so we can rewrite Eq. (2.3) as

$$\frac{\partial^2 (\psi_1 + \psi_2)}{\partial x^2} - \frac{1}{c^2}\frac{\partial^2 (\psi_1 + \psi_2)}{\partial t^2} = 0. \tag{2.4}$$

Thus, the sum of the two waves is a solution to the wave equation.

Now let's refer back to Fig. 2.1 and examine the principle of superposition for waves of the form $E(x,t) = E_0 e^{i(kx-\omega t)}$. The waves pictured have the same wavelength and amplitude but differ by a constant phase shift, i.e. are coherent. Using the principle of superposition we can find the electric field produced by the two overlapping waves;

$$E(x,t) = E_0 e^{i(kx-\omega t)} + E_0 e^{i(kx-\omega t+\phi)}. \tag{2.5}$$

We will try and simplify this expression to express the resultant wave as simply as possible. First, we will pull out a common factor;

$$E(x,t) = E_0 e^{i(kx-\omega t)}\left[1 + e^{i\phi}\right]. \tag{2.6}$$

This we can express as

$$E(x,t) = E_0 e^{i(kx-\omega t+\phi/2)}\left[e^{-i\phi/2} + e^{i\phi/2}\right]. \tag{2.7}$$

Now we will rewrite the expressions in the brackets in terms of a cosine function;

$$E(x,t) = 2\cos(\phi/2) E_0 e^{i(kx-\omega t+\phi/2)}. \tag{2.8}$$

This expression is quite illuminating in that it allows us to see the way in which the amplitude of the resultant depends on the phase shift. The maximum absolute amplitude occurs when $\cos(\phi/2) = \pm 1$, for which ϕ is an integer multiple of 2π, i.e. $\phi = m2\pi$, $m = 0, 1, 2, ...$ and is twice the amplitude of each of the original waves. This is called constructive interference. When $\phi = 2\pi(m+1/2)$ the resultant electric field is zero everywhere. This is called destructive interference. As we vary the angle from 0 to 2π we obtain a range of the absolute amplitude between 0 and $2E_0$. The waves in the Fig. 2.1 are shifted by an arbitrary phase and give a feeling for how the resultant wave depends on the shift. Superposition can also be applied to waves moving in the same direction but having different frequencies.

2.2 – Superposition in the Complex Plane

Another mathematical approach for superposing the waves in Eq. (2.5) is graphical, as illustrated in Fig. 2.2. Each of the complex waves is represented as a vector in the complex plane. One is represented in red at angle $kx - \omega t$ from the real axis, and the other in green at angle $kx - \omega t + \phi$ from the real axis. The resultant, or sum, in blue bisects these two vectors since they are equal in amplitude, and therefore is at angle $\phi/2$ from both. The resultant therefore has a length $2E_0 \cos(\phi/2)$, at an angle $kx - \omega t + \phi/2$ from the real axis; thereby making its real projection $2E_0 \cos(\phi/2) \cos(kx - \omega t + \phi/2)$, in agreement with Eq. (2.8). As time progresses, for a given position x, the resultant vector maintains a constant amplitude $2E_0 \cos(\phi/2)$ while rotating clockwise. Any number of waves can be superposed by this geometrical method.

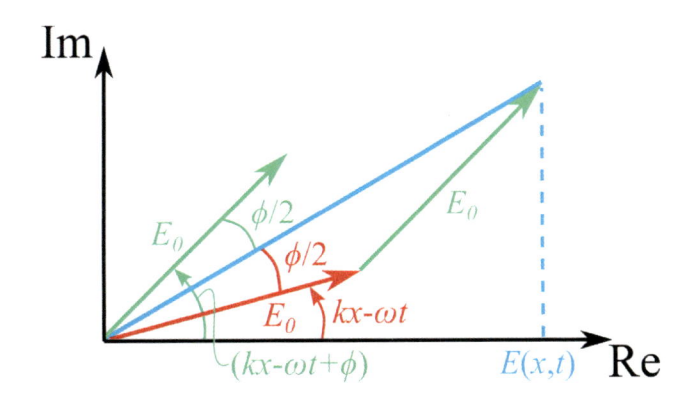

Figure 2.2 – Graphical method for adding waves in the complex plane.

2.3 – Constructing a Holographic Diffraction Grating

Now we will examine how the principle of superposition is used to create a diffraction grating. A holographic diffraction grating can be manufactured by superimposing plane waves of light on a special high-resolution photographic film. Before the laser, a typical grating was generated by scribing thousands of parallel lines in a glass plate with a tool, a time consuming and wearing process, since the pitch, or separation between lines, was on the order of one millionth of a meter (1 micron). With such a pitch a typical grating has 1000 lines per millimeter so a one-inch wide grating would require about 25,000 lines to be cut into the glass plate. This approach was changed by the apparatus in Fig. 2.3, which produces a diffraction grating with a single pulse of a laser.

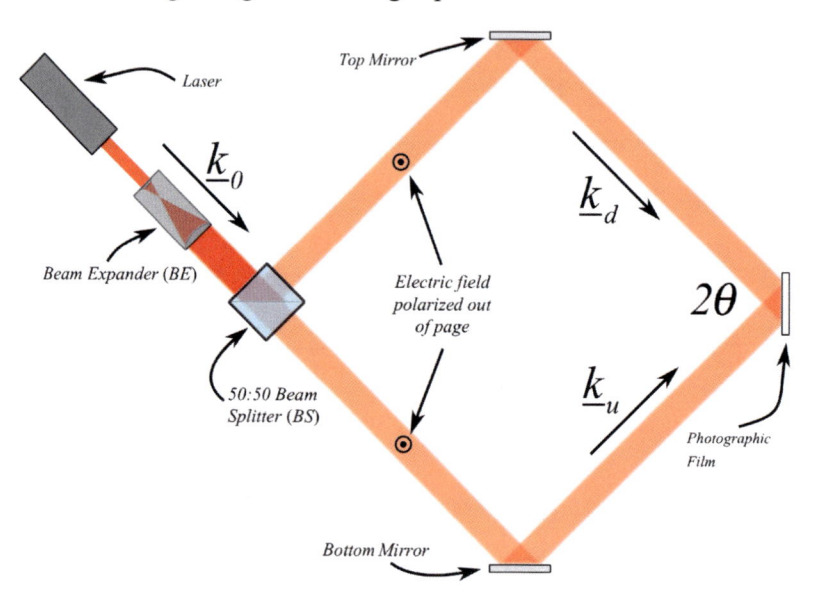

Figure 2.3 – How to construct a holographic diffraction grating.

The developed photographic film is known as holographic diffraction grating. The name will become clear after we complete the analysis below. It is generated by expanding a laser beam using a dual lens beam expander (BE), followed by splitting the wave front with a beam splitter (BS) that generates two beams, one which proceeds to the mirror at the bottom and the other to the mirror at the top. Reflection from these mirrors generates beams moving upward from the lower mirror and downward from the upper mirror. These beams overlap and interfere on the photographic film, Fig. 2.4. We can use the principle of superposition to describe the interference pattern produced on the photographic film.

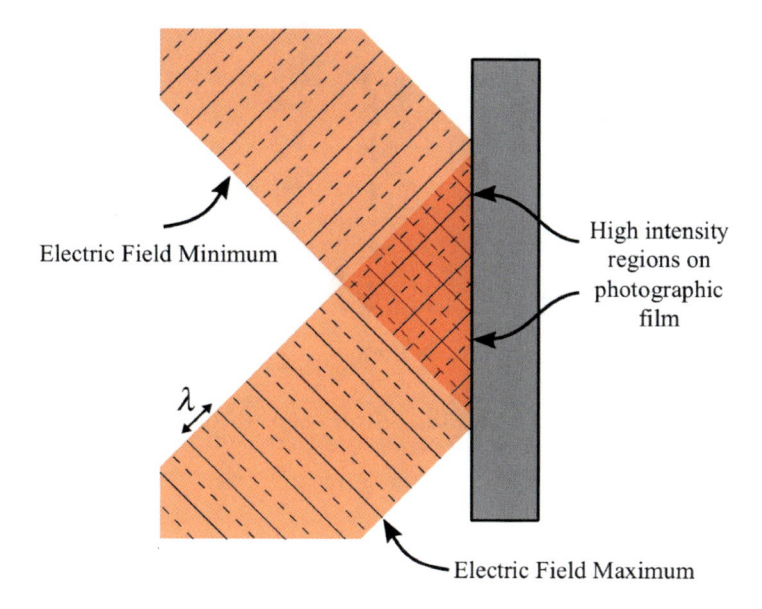

Figure 2.4 – Interference of EM waves on photographic film. Maxima in intensity recorded on the photographic film occur with field maxima overlap or field minima overlap.

The total electric field incident on the film plane is

$$\underline{E}_G(\underline{r},t) = E_0 \hat{z}\left[e^{i(\underline{k}_u \cdot \underline{r} - \omega t)} + e^{i(\underline{k}_d \cdot \underline{r} - \omega t)} \right]. \quad (2.9)$$

Here the field has been split into upward and downward beams, each of equal amplitude and polarization. The intensity of the light incident on the film is

$$\langle I(\underline{r}) \rangle_t = \frac{1}{2}\varepsilon_0 c \left[\underline{E}_G \cdot \underline{E}_G^* \right], \text{ or} \quad (2.10)$$

$$\langle I(\underline{r}) \rangle_t = \frac{1}{2}\varepsilon_0 c E_0^2 \left[e^{i(\underline{k}_u \cdot \underline{r} - \omega t)} + e^{i(\underline{k}_d \cdot \underline{r} - \omega t)} \right] \cdot \left[e^{-i(\underline{k}_u \cdot \underline{r} - \omega t)} + e^{-i(\underline{k}_d \cdot \underline{r} - \omega t)} \right], \quad (2.11)$$

which simplifies to

$$\langle I(\underline{r}) \rangle_t = \frac{1}{2}\varepsilon_0 c E_0^2 \left[2 + e^{i(\underline{k}_u - \underline{k}_d)\cdot \underline{r}} + e^{-i(\underline{k}_u - \underline{k}_d)\cdot \underline{r}} \right], \quad (2.12)$$

or

$$I(\underline{r}) = 2\varepsilon_0 c E_0^2 \cos^2\left[\frac{(\underline{k}_u - \underline{k}_d)}{2} \cdot \underline{r} \right]. \quad (2.13)$$

So it appears the intensity on the screen varies as cosine squared, however the vector form in the argument is not quite as clear as we may like. Let's examine the vectors involved, see Fig. 2.5.

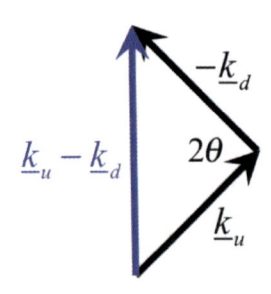

Figure 2.5 – Resultant of two wave vectors.

We know that the vectors have the same magnitude, i.e. $k_u = k_d = k$. Using some basic trigonometry we can rewrite the sum as

$$\underline{k}_u - \underline{k}_d = 2k\sin(\theta)\hat{y}. \tag{2.14}$$

The coordinate system we have chosen has its origin at the center of the photographic film so $\underline{r} = y\hat{y}$ as the grating is along the y-axis. We can then rewrite the terms inside the cosine function in Eq. (2.13) as

$$\left(\underline{k}_u - \underline{k}_d\right)\cdot\underline{r} = 2k\sin(\theta)y. \tag{2.15}$$

So that the intensity from Eq. (2.13) is

$$\langle I(y)\rangle_t = 2\varepsilon_0 c E_0^2 \cos^2\left[k\sin(\theta)y\right]. \tag{2.16}$$

A plot of the intensity pattern produced by the cosine-squared expression is shown in Fig. 2.6.

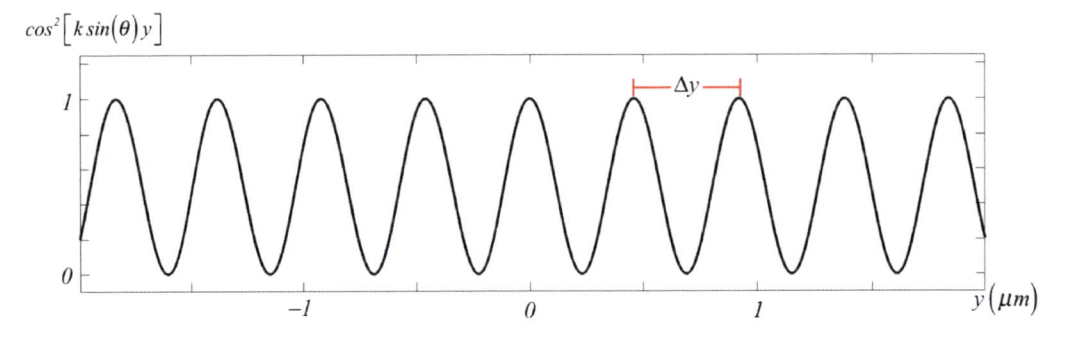

Figure 2.6 – Cosine squared intensity diffraction pattern produced for $\lambda = 600\,nm$ and $\theta = 45°$.

The photographic film in Fig. 2.3 is assumed to record in proportion to the intensity incident on it. A diffraction grating of this form produces three diffracted beams of high intensity for laser beam incident perpendicular to its plane. We will learn why this pattern is produced when we study diffraction. Now it is interesting to compute the expected spacing between the maxima (pitch).

Recalling that $k = \frac{2\pi}{\lambda}$, the line spacing (a.k.a. pitch) that is produced as a function of angle θ in Fig. 2.2 is

$$\Delta y = \frac{\lambda}{2\sin(\theta)}. \qquad (2.17)$$

This spacing is obtained by examining Eq. (2.16).

Suppose we use a wavelength $\lambda = 600$ *nm* and $\theta = 45°$ then the spacing is

$$\Delta y = \frac{600 \; nm}{\sqrt{2}} = 424 \; nm. \qquad (2.18)$$

A standard grating may have a pitch of one micron, or 1000 nm. If we suppose such a grating was produced by a laser with a wavelength of 650 nm, then what would be the angle θ (Exercise 2.1)?

Shining a laser beam on this type grating only produces three beams corresponding to two first-order beams and a 0th-order beam. This is rather odd since introductory physics texts often describe diffraction gratings as producing a larger number of maxima, for which the order can have any value, so long as the diffracted angle does not exceed 90°. The reason for this perceived anomaly will be discussed in our study of diffraction.

There is a reason that the grating constructed in Fig. 2.3 is called holographic. It has to do with wave reconstruction. To understand where such reconstruction appears, we will first describe the processing of the film. The film is processed by taking it into a darkroom and applying chemicals (developer) in order to produce a negative associated with the exposure. The negative darkens those areas that are exposed to light by the nucleation of metallic colloidal particles of silver. When the negative is returned to the place from which the film was irradiated the light interaction produces a rather strange and wonderful effect. To realize this effect we will have to make a slight change in the optical configuration in Fig. 2.3. A shutter will be put into the path of the beam that is moving downward toward the original film; the developed film is only exposed to the upward beam. Now with our eyes we gaze at the back of the developed film. Amazingly, the downward beam reappears in our eyes. In other words we have reconstructed the wave front that was originally there, even though the downward beam has been blocked. The film is now called a hologram; it records information in three dimensions. Holography, the subject built around this way of storing 3D information, was invented by Dennis Gabor between 1946 and 1951, for which he received the Nobel Prize in Physics in 1971. When we begin the study of the diffraction of light we will return to the holographic diffraction grating and prove theoretically that such beams reappear. Had Princess Leia been seated in place of the top mirror, light scattered from her superimposed with our up beam would have constructed her hologram as was demonstrated in Star Wars – A New Hope (1977).

2.4 – Diffraction from a Pinhole

We now will deal with the way in which light interacts with obstacles. A pupil is an example. It represents a clear opening in an otherwise opaque iris. An atom scatters light and therefore is an obstacle. Different parts of the obstacle can scatter light, which interfere with each other through the principle of superposition. In this section we are interested in analyzing such scattering to determine the geometry of structures, (e.g. diffraction gratings, DNA). We will start off by describing a simple example and work our way all the way up to the scattering from a single molecule, and DNA.

Imagine you have a plane wave incident on a pinhole that is much smaller than the wavelength, Fig. 2.7.

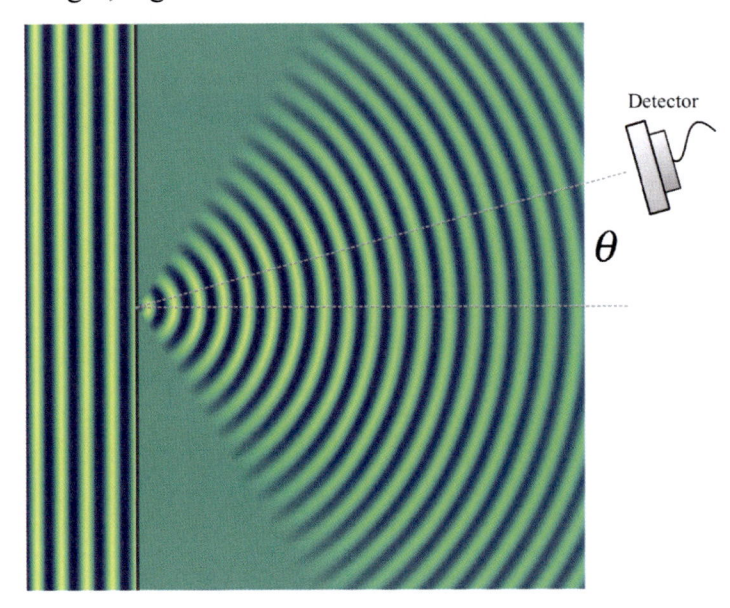

Figure 2.7 – A plane wave diffracting through a single pinhole aperture.

When the plane wave reaches this "aperture" the wave will enter the hole and spread upon leaving, a phenomenon known as diffraction. To measure how the light scatters we can position a detector at an arbitrary angle θ relative to the incident wave. The configuration is shown in Fig. 2.7. The detector measures the intensity of the light incident on it. As we vary the position of the detector we are able to measure how the intensity changes as a function of the scattering angle θ.

2.5 – Diffraction and the Fourier Transform

Now instead of working with a single point let's consider a slit aperture with some width and see how light will diffract when passing through the slit.

Huygen's principle states that every point along the wave front within the slit acts as a point source of light known as a wavelet. The result of representing a plane wavefront entering a slit by a large number of wavelets is shown in Fig. 2.8. As one gets several wavelengths from the slit plane, the superposition of these waves shows distinct spreading along with an angular pattern of intensity. On the screen on the right side of the figure the recorded intensity has a prominent central peak along with weaker secondary peaks as the off-axis angle increases.

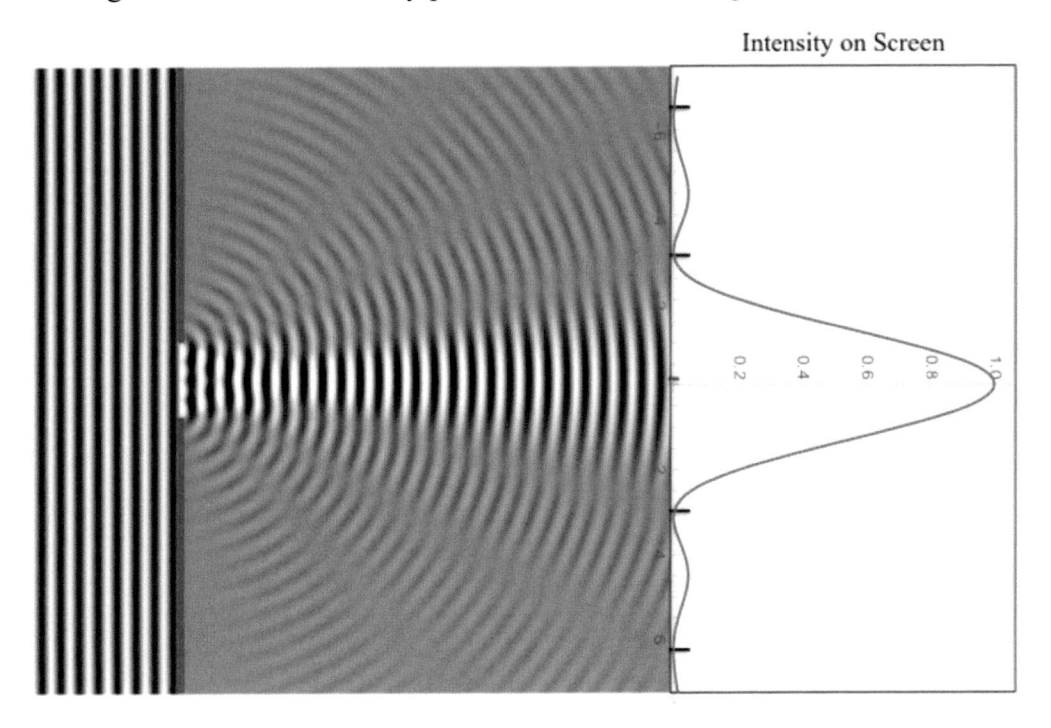

Figure 2.8 – Diffraction through a single slit 4 wavelengths wide. The intensity on the screen 26 wavelengths from the slit plane is plotted on the right side. The simulation on the left was computed (Richard F. Lyon via Wikimedia Commons).

51

The result illustrated on the screen in Fig. 2.8 is actually the pattern that forms a large distance away, in the so-called Fraunhofer region. Before calculating the pattern due to an infinite number of wavelets, we will qualitatively describe what is seen by placing the screen at different distances from the slit.

Had we pressed the screen against the slit we would not see the spatial intensity plot on the right side of Fig. 2.8 with its multiple peaks and zeros. Rather, we would observe the shape of the aperture (i.e. slit). Indeed we see in Fig. 2.8 a lack of speading a few wavelengths from the slit. As the screen is moved away the shape of the image on the screen will continuously change. This is known as the region of Fresnel, or near-field diffraction. At a larger distance the diffraction image stabilizes, and we are in the region of Fraunhofer, or far-field diffraction. That's what is depicted by the intensity pattern in Figure 2.8, which is recorded 26 wavelengths from the slit plane. If the aperture width is a, and the distance between the the aperture plane and the screen plane is D, then a rule of thumb is that the diffraction will have the stable Fraunhofer image when $D > a^2/\lambda$. Our major interest going forward will be in Fraunhofer diffraction. With the theory we develop to explain far-field diffraction we will decipher a host of diffraction images such as the screen image in Fig. 2.8, the image caused by a diffraction grating, and the x-ray diffraction image of DNA.

Fig. 2.9 is a schematic from which we will attempt to understand the intensity pattern in Fig. 2.8. The figure is massively exaggerated in the relative scale of the slit in relation to the distance to the screen; to be in the Franhofer regime, considering the wavelength shown, R should be much larger than the slit width a or wavelength λ.

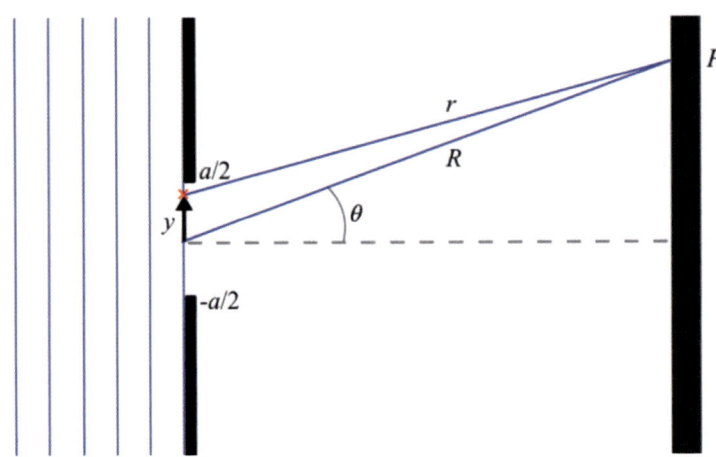

Figure 2.9 – Geometrical description of diffraction.

As a wave enters the slit, it is assumed to have a plane wave front (i.e. flat). Each point in this slit emits as if it were a strip of infinitesimal width; the strip is very long into the page. We would like to know what the intensity on the screen looks like, and I have picked an arbitrary point P at which I would like to determine this intensity. If the width of the emitting source at height y is dy and the field strength from this differential source is $\underline{\mathcal{E}}\,dy$ then the field at P due to the source at y is approximately

$$d\underline{E}_p = \underline{\mathcal{E}} e^{i(kr-\omega t)} dy . \tag{2.19}$$

Note that I have ignored the attenuation of the wave amplitude with the distance travelled r, since r does not vary much from one differential source to another, so this attenuation varies slowly, roughly as that from a cylindrical source, $1/r^{1/2}$. However the phase kr can vary considerably since it has a 2π cycle, and k can be very large. For this reason it is essential to find out how r in the exponent changes with y. Fortunately, when the distance to the screen is much larger than the slit width (i.e. $y/R \ll 1$), it is possible to express r to first order in y, for a given R and angle θ. In short

$$r \simeq R[1 - \frac{y}{R}\sin(\theta)] = R - y\sin(\theta) . \tag{2.20}$$

You can confirm this by writing down the exact relationship for r as a function of y, R, and θ by applying the law of cosines to the associated triangle in Fig.2.9, and expanding what you get in terms of powers of y/R (see Appendix A.1). Eq. (2.20) is what would be expected if the path marked r in Fig. 2.9 were parallel to the central path R.

With r from Eq. (2.20) substituted into Eq. (2.19), the field from the differential source becomes

$$d\underline{E}_P = \underline{\mathcal{E}} e^{i(kR-\omega t)} e^{-ik\sin(\theta)y} dy . \tag{2.21}$$

Now it is only a matter of integrating over all the differential sources to find the field at P;

$$\underline{E}_P = \underline{\mathcal{E}} e^{i(kR-\omega t)} \int_{-a/2}^{a/2} e^{-ik\sin(\theta)y} dy . \tag{2.22}$$

53

We are interested only in the intensity, $I_P \propto E_P^* E_P$;

$$I_p \propto \mathcal{E}^2 \left| \int_{-a/2}^{a/2} e^{-ik\sin(\theta)y} \, dy \right|^2. \tag{2.23}$$

It is convenient to identify $k\sin(\theta)$ as the y component of the diffracted wave vector, k_y. This trims our result to

$$I_p \propto \mathcal{E}^2 \left| \int_{-a/2}^{a/2} e^{-ik_y y} \, dy \right|^2. \tag{2.24}$$

Note that the limits of the integral include only the regions where the aperture is transparent. If we extend the integration from $-\infty$ to ∞ we can introduce an "aperture function" to carve out this transparent region. Then the integral in Eq. (2.24) becomes

$$c(k_y) = \int_{-\infty}^{\infty} A(y) e^{-ik_y y} \, dy. \tag{2.25}$$

The form of this integral is called a "Fourier Transform" of $A(y)$. The intensity on the screen is just proportional to the square modulus of the Fourier transform of the aperture function $A(y)$. As a consequence of integrating over y the integral will only be a function of k_y. For our particular problem

$$A(y) = \begin{cases} 1, & -\dfrac{a}{2} < y < \dfrac{a}{2} \\ 0, & otherwise \end{cases}. \tag{2.26}$$

This is shown in Fig. 2.10.

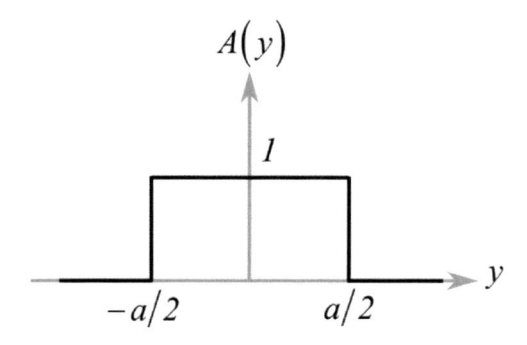

Figure 2.10 – A single slit aperture function of width a.

In general $A(y)$ can have any pattern and can be complex if for example a refractive material occupies the aperture. Over time we will discuss a host of aperture functions. For now we might as well finish the discussion of the slit and see how well our result agrees with Fig. 2.8. This is a matter of applying Eq. (2.25) to the aperture function in Fig. 2.10,

$$c(k_y) = \int_{-\infty}^{\infty} A(y)e^{-ik_y y}\, dy = \int_{-a/2}^{a/2} e^{-ik_y y}\, dy. \qquad (2.27)$$

The integral on the far right in Eq. (2.27) can be evaluated by remembering that it should produce an exponential. After a bit of algebra we get

$$c(k_y) = \left[\frac{e^{-ik_y y}}{-ik_y}\right]_{-a/2}^{a/2} = \frac{e^{-ik_y a/2} - e^{ik_y a/2}}{-ik_y} = \frac{a \sin(k_y a/2)}{k_y a/2}. \qquad (2.28)$$

Substituting the right most result into Eq. (2.24) shows that the intensity is

$$I \propto \mathcal{E}^2 a^2 \left[\frac{\sin(k_y a/2)}{k_y a/2}\right]^2 = \mathcal{E}^2 a^2 \operatorname{sinc}^2(k_y a/2), \qquad (2.29)$$

where the function $\operatorname{sinc}(x) = \sin(x)/x$. The argument $k_y a/2$ increases with the vertical position along the screen in Fig. 2.9. That is because k_y is the y-component of the wave vector $\underline{k}$ describing the propagation from the center of the slit to point P on the screen; this y-component increases as we move up on the screen. Consequently we want the k_y dependence of Eq. (2.29) or equivalently the dependence on $k_y a/2$. The intensity pattern is shown in Fig. 2.11. You can use Wolfram Alpha, or another graphing tool, to display this result.

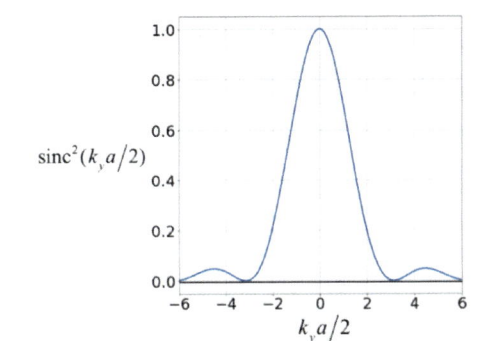

Figure 2.11 – *The intensity of light diffracted through a single slit varies as a sinc-squared function.*

55

Fig. 2.11 has the same shape as the intensity plot in Fig. 2.8. There are interesting effects associated with Eq. (2.29). One of these is that the width of the central peak on the screen is reduced as the slit width is enlarged, an effect which is counter intuitive. The general principle arrived at from our work is that the diffraction display on the screen is described by the square modulus of the Fourier transform, Eq. (2.29).

The Fourier transform is used to calculate an intensity pattern created by an aperture

$$c(k_y) = \int_{-\infty}^{\infty} A(y) e^{-ik_y y} dy. \qquad (2.30)$$

It is also possible to find an aperture function from a field pattern. This is called an inverse Fourier transform

$$A(y) = \frac{1}{2\pi} \int_{-\infty}^{\infty} c(k_y) e^{ik_y y} dk_y. \qquad (2.31)$$

The inverse Fourier transform allows us to determine the shape of an aperture if we have a formula for the scattered field. This makes it possible to determine the structure of something too small to be imaged with a light microscope. For example, DNA is much too small to be imaged with a light microscope; its structure was first experimentally determined using X-ray diffraction. Going forward we will need to model apertures of chemical structures in order to determine how light diffracts from them. This will require modeling points in space.

2.6 – Dirac Delta function

Aperture functions for structures that are infinitesmal in one, two or three dimensions can be easily modeled using a mathematical device called the delta function, since it can represent singularities in one, two and three dimensions. I will start first with the one dimensional case (e.g. infinitesimally narrow slit).

The delta function is a spike to infinity at the point where its argument is zero, and zero everywhere else. The function is written as

$$f(y) = \delta(y - y_1). \qquad (2.32)$$

The delta function is presented graphically in Fig. 2.12. The red line along the y-axis represents the delta function. It is a spike existing at $y = y_1$.

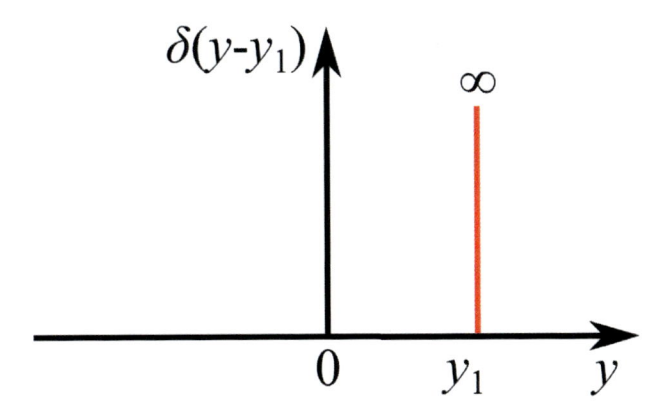

Figure 2.12 – Graph of a delta function $\delta(y - y_1)$.

The delta function has infinitesimal width and infinite height such that the area under the graph is 1;

$$\int_{-\infty}^{\infty} \delta(y - y_1) \, dy = 1. \tag{2.33}$$

The delta function has a sampling property when multiplied by a continuous function in an integral; it picks out the value of this function at the point at which the delta function is singular. This is formally expressed by the following equation:

$$\int_{-\infty}^{\infty} f(y) \delta(y - y_1) \, dy = f(y_1). \tag{2.34}$$

In diffraction problems the delta function makes a very practical aperture function. The point created can be used as a 1D hole (infinitesimally narrow slit). Imagine that $\delta(y - y_1)$ is a small opening in the y-axis located at $y = y_1$. The diffraction through such an aperture spreads to all k_y values. You can readily see this in evaluating the diffracted field, proportional to $c(k_y)$ in Eq. (2.30), for this aperture,

$$\int_{-\infty}^{\infty} \delta(y - y_1) e^{-ik_y y} \, dy = e^{-ik_y y_1}. \tag{2.35}$$

The result has no restriction on k_y, furthermore its square modulus is simply 1, independent of k_y.

57

The inverse transform of the result in Eq. (2.35) provides a representation for the delta function since the inverse must give the aperture function, and in this case $A(y) = \delta(y - y_1)$. So applying the inverse transform [Eq. (2.31)] to $c(k_y) = \exp(-ik_y y_1)$ gives

$$\delta(y - y_1) = \frac{1}{2\pi} \int_{-\infty}^{\infty} e^{-ik_y y_1} e^{ik_y y} dk_y = \frac{1}{2\pi} \int_{-\infty}^{\infty} e^{ik_y (y - y_1)} dk_y. \quad (2.36)$$

This is now a useful representation for the delta function. It is important to note that the delta function is even; i.e. $\delta(y - y_1) = \delta(y_1 - y)$.

The delta function can be extended to represent apertures in multiple dimensions or to represent familiar apertures such as Young's double slit. The double slit can be modeled as the sum of two delta functions located at $y = d/2$ and $y = -d/2$;

$$A(y) = a\left[\delta(y - d/2) + \delta(y + d/2)\right]. \quad (2.37)$$

The diffraction pattern generated by these can be found via a Fourier Transform of Eq. (2.37);

$$c(k_y) = a\int_{-\infty}^{\infty}\left[\delta\left(y - \frac{d}{2}\right) + \delta\left(y + \frac{d}{2}\right)\right] e^{-ik_y y} dy = a\left(e^{-i\frac{k_y d}{2}} + e^{+i\frac{k_y d}{2}}\right) = 2a\cos\left(\frac{k_y d}{2}\right) \quad (2.38)$$

If you plot out the intensity $|c(k_y)|^2$ you will see that it has a form that you would expect for diffraction from Young's double slits. In particular there will be maxima in intensity when $k_y d/2 = m\pi$, where m is a positive or negative integer known as the order of the diffraction peak. Since $k_y = (2\pi/\lambda)\sin(\theta)$, the angles at which the maxima are found correspond to

$$\sin(\theta_m) = m\lambda/d, \quad (2.39)$$

a result that should be familiar from your introductory physics course.

The delta function for a hole in an aperture in 2D is given by $\delta(x, y)$. This 2D delta function is the product of two 1D delta functions, one in each of the chosen coordinates;

$$\delta(x, y) = \delta(x - x_1)\delta(y - y_1). \quad (2.40)$$

As the 1D delta function is a spike where its argument is zero and represents a hole in a 1D aperture, the 2D delta function is a hole in a 2D aperture at the coordinate (x_1, y_1). The 2D delta function becomes extremely useful in modeling individual atoms in a plane, as you will see in Exercises 2.7 and 2.10.

The delta function can be extended still further to model points in 3D space, the function for this is

$$\delta(x,y,z) = \delta(x-x_1)\delta(y-y_1)\delta(z-z_1). \qquad (2.41)$$

Eq. (2.41) represents a point at coordinates (x_1, y_1, z_1).

Before we continue analyzing diffraction patterns let's discuss an interesting fact about diffraction. Imagine we have two aperture functions such that they are perfect complements. By complements we mean that one aperture is opaque while the other is transparent. An example of complementary aperture functions is shown in Fig. 2.13.

Figure 2.13 – Complementary apertures.

Complementary aperture functions produce the same intensity patterns. This is known as Babinet's principle. The waves diffracted off complementary apertures are opposite in phase but the intensity pattern produced on a distant screen is the same shape. This is a very fortunate result and will greatly simplify our study of diffraction from molecular structures such as DNA; we can diffract off ultra-small holes in an aperture, but that is equivalent in terms of the diffracted intensity to diffracting off isolated points.

2.7 – Fraunhofer Diffraction in Multiple Dimensions

We have previously considered diffraction caused by an aperture with two adjacent openings, such as Young's double-slit, in the case that light is incident normal to the position vector between the two slits. In these problems the plane wave of light hits both slits with the same phase. If we are comfortable calculating the diffraction pattern produced in such a case it is interesting and useful to generalize the experiment for incident light that is not perpendicular to the slit plane but rather approaches at an arbitrary angle.

To start, we will construct the aperture in Fig. 2.14 consisting of two atoms. Recall Babinet's principle, which states that the intensity pattern is the same whether we use holes or atoms. Because light is incident at an arbitrary angle in general, the light will have different phases at a given time as it irradiates each feature. Both the upper and lower points diffract the light into radial waves that will interfere like the ripples in a pond that form when multiple splashes are made. It seems natural that the interference pattern should depend on the angle at which the light is incident relative to the inter-point vector $\underline{r}$.

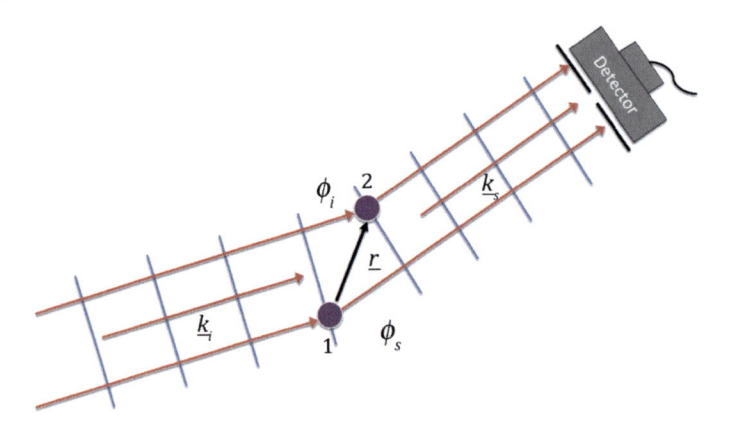

Figure 2.14 – Scattering from two points in space.

Fig. 2.14 is drawn for a particular time $(t = 0)$, thereby eliminating a temporal phase. The upper point is displaced by $\underline{r}$ from the lower point, and differential incident phase extension beyond the lower point is $\phi_i = \underline{k}_i \cdot \underline{r}$. Beyond this the upper point scatters the wave, which has a shorter distance to the detector than the lower point. The additional distance from the lower point amounts to an additional scattered phase of $\phi_s = \underline{k}_s \cdot \underline{r}$. The overall difference in these phases is $\phi_s - \phi_i$. If each point has unit scattering amplitude, then the amplitude measured at the detector is

$$c(\underline{k}_s - \underline{k}_i) = 1 + e^{-i(\underline{k}_s - \underline{k}_i) \cdot \underline{r}} . \tag{2.42}$$

For a more complicated aperture $A(y)$ we will have to add up the scattering amplitude from all points in the aperture. This naturally produces a 3-D Fourier transform :

$$c(\underline{k}_s - \underline{k}_i) = \int_{Vol} A(\underline{r}) e^{-i(\underline{k}_s - \underline{k}_i) \cdot \underline{r}} d^3\underline{r}. \qquad (2.43)$$

To make things a little neater we substitute $\underline{q} = \underline{k}_s - \underline{k}_i$ into the equation. Returning to Eq. (2.43) we find:

$$c(\underline{q}) = \int_{Vol} A(\underline{r}) e^{-i\underline{q} \cdot \underline{r}} d^3\underline{r}. \qquad (2.44)$$

As an example of the use of Eq. (2.44) we will return to the holographic diffraction grating. As you recall, we said that if we kept the upward beam after installing the developed film, but did away with the light reflected from the top mirror, then by viewing the field of light diffracting from the film, the decending beam would be reconstructed. As we go forward we will first discuss diffraction from the grating for perpendicular incidence, before returning to holographic wave reconstruction.

Eq. (2.44) will allow us to tackle all of the diffraction problems we will encounter in this text, however, it is important that we recognize its limitations. Our approach will not allow us to describe light that scatters many times on its way to the detector. There are some apertures in which light will scatter at a point and then scatter again before reaching the detector. This is called multiple scattering. In this text we only consider apertures such that the effects of multiple scattering are negligible.

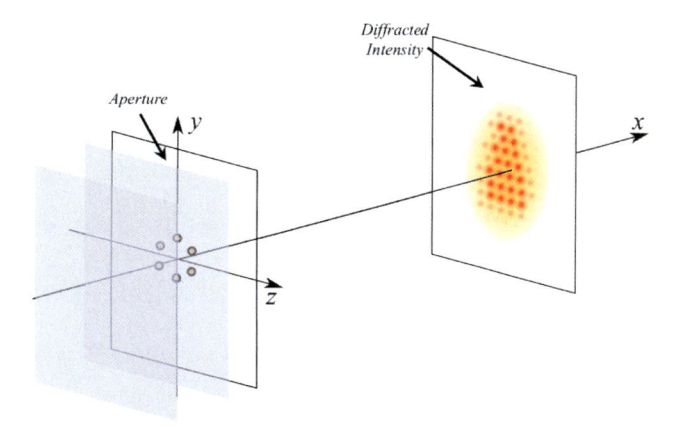

Figure 2.15 – Plane waves diffracting off an aperture confined to yz-plane.

Let's consider a special case where the aperture is confined to the *yz*-plane, Fig. 2.15. In this case the vector $\underline{r}$ in Eq. (2.44) can be written in terms of y and z components. The Fourier transform for such an aperture is

$$c(k_y, k_z) = \int\int A(y,z) e^{-i\underline{q}\cdot(y\hat{y}+z\hat{z})} \, dy\, dz . \qquad (2.45)$$

The exponent is easy to evaluate as the dot product of two vectors. Let's further consider that the incident wave is along the x-axis; i.e. $\underline{k}_i \cdot \underline{r} = 0$, as it would be in the benzene problem you were assigned (Exercise 2.7). So $\underline{q} = k_{sy}\hat{y} + k_{sz}\hat{z}$, allowing us to write Eq. (2.45) as

$$c(k_{sy}, k_{sz}) = \int\int A(y,z) e^{-i(k_{sy}y + k_{sz}z)} \, dy\, dz . \qquad (2.46)$$

The intensity of the scattered light varies as

$$I \propto \left| c(k_{sy}, k_{sz}) \right|^2 . \qquad (2.47)$$

Now that we have developed Eq. (2.46) for describing far-field diffraction with incident light perpendicular to the aperture plane we would like to return to holographic grating we discussed at the beginning of the chapter and determine how light scatters through it at an arbitrary angle. The cosine-squared diffraction grating described in Section 2.2 scatters the light that travels through it in an unusual way. This grating scatters a single beam of light from a laser into three distinct maxima. In order to understand this we will have to recall how the grating was constructed. The grating was constructed with a light intensity [Eq. (2.16)],

$$\langle I(y) \rangle_t = 2\varepsilon_0 c E_0^2 \cos^2\left[k\sin(\theta)y\right].$$

This intensity pattern relates to the aperture function of the grating formed by it. The aperture function is

$$A(y) = \cos^2\left(\frac{\pi}{P_s} y\right), \qquad (2.48)$$

where P_s is the spatial period (a.k.a. pitch) or spacing between maxima. A readily available cosine-squared diffraction grating has 1000 lines per millimeter, or a pitch of $1\,\mu m$. In order to determine the intensity pattern produced by light scattered through this aperture we take the Fourier transform, which in effect assumes that the beam irradiates an infinite number of slits[9];

$$c(k_y) = \int_{-\infty}^{\infty} \cos^2\left(\frac{\pi}{P_s}y\right) e^{-ik_y y} dy. \quad (2.49)$$

To evaluate this we can use the trigonometric identity $\cos^2(\theta) = \frac{1}{2} + \frac{1}{2}\cos(2\theta)$ to rewrite the aperture function,

$$c(k_y) = \int_{-\infty}^{\infty} \left[\frac{1}{2} + \frac{1}{2}\cos\left(\frac{2\pi}{P_s}y\right)\right] e^{-ik_y y} dy. \quad (2.50)$$

The cosine function can be expressed in terms of complex exponentials;

$$c(k_y) = \int_{-\infty}^{\infty} \left[\frac{1}{2} + \frac{1}{4}\left(e^{i\frac{2\pi}{P_s}y} + e^{-i\frac{2\pi}{P_s}y}\right)\right] e^{-ik_y y} dy. \quad (2.51)$$

Now the integral has three terms, which we can integrate over individually

$$c(k_y) = \frac{1}{2}\int_{-\infty}^{\infty} e^{-ik_y y} dy + \frac{1}{4}\int_{-\infty}^{\infty} e^{i\left(\frac{2\pi}{P_s} - k_y\right)y} dy + \frac{1}{4}\int_{-\infty}^{\infty} e^{-i\left(\frac{2\pi}{P_s} + k_y\right)y} dy. \quad (2.52)$$

These integrals evaluate to delta functions;

$$c(k_y) = \pi\delta(k_y) + \frac{\pi}{2}\delta\left(\frac{2\pi}{P_s} - k_y\right) + \frac{\pi}{2}\delta\left(\frac{2\pi}{P_s} + k_y\right). \quad (2.53)$$

The delta functions correspond to the three diffracted beams we observed when we shined the laser at normal incidence on the holographic diffraction grating in the Pocket Optics kit! The positions of the maxima are at $k_y = 0, \pm 2\pi/P_s$. The center beam ($k_y = 0$) has twice the field amplitude of the side beams ($k_y = \pm 2\pi/P_s$). As in the Young's double slit experiment, the angle of each of these beams correspond to $\sin(\theta_{+1,-1}) = \pm\lambda/P_s$. The major difference here is that there are only 3 diffraction maxima. Next we use Eq. (2.45) to analyze the

---

[9] A laser beam is not infinite in width, however our assumption simplifies the mathematics. In Exercise 2.5 the laser beam width is reduced to a finite dimension L.

diffraction associated with the oblique incidence on the same grating from the upward directed beam in Fig. 2.3

To begin let's recall the aperture function used to represent the grating; $A(\underline{r}) = \cos^2[k \cdot \sin(\theta)y]\delta(x)$. Here we have supposed that the grating is very thin. In this case we have chosen to align the grating "slits" along the z-axis as before. This time, however, the incident light is not normal to the surface. Instead we can have the light incident on the surface at an angle θ from the grating normal as in Fig. 2.3. To determine how light scatters from this aperture we can apply the generalized transform;

$$c(\underline{q}) = \int_{Vol} A(\underline{r})e^{-i\underline{q}\cdot\underline{r}}d^3\underline{r} = \int_{Vol} \cos^2\left(\frac{k_g y}{2}\right)\delta(x)e^{-i\underline{q}\cdot\underline{r}}d^3\underline{r}. \quad (2.54)$$

Since the grating "slits" are separated along the y-axis, it is convenient to reduce the volume integration to integrating along y, by taking $\underline{r} = y\hat{y}$,[10]

$$c(\underline{q}) = \int \cos^2\left(\frac{k_g y}{2}\right)e^{-i(k_{sy}-k_{iy})y}dy \text{, or} \quad (2.55)$$

$$c(\underline{q}) = \int \left[\frac{1}{2}+\frac{1}{2}\cos(k_g y)\right]e^{-i(k_{sy}-k_{iy})y}dy. \quad (2.56)$$

In exponential form this is

$$c(\underline{q}) = \int \left[\frac{1}{2}+\frac{1}{4}\left(e^{ik_g y}+e^{-ik_g y}\right)\right]e^{-i(k_{sy}-k_{iy})y}dy. \quad (2.57)$$

Evaluating the integral in the simplest possible way by extending the integration limits from $-\infty$ to $+\infty$, we get

$$c(\underline{q}) = \pi\delta(k_{iy}-k_{sy})+\frac{\pi}{2}\left[\delta(k_{iy}-k_{sy}+k_g)+\delta(k_{iy}-k_{sy}-k_g)\right]. \quad (2.58)$$

A special case arises when the incident beam is shined at the same angle as the beam used to produce the grating, i.e. $k_{iy} = k\sin\theta$. Since $k_g = 2k\sin\theta$ using Eq. (2.17), $k_g = 2k_{iy}$. Returning to Eq. (2.58), we see that it can be written as

---

[10] I have left out an arbitrary constant associated with the width of the physical system in the y and z – directions; a plane wave approximation.

$$c(\underline{q}) = \pi\delta(k_{iy} - k_{sy}) + \frac{\pi}{2}\left[\delta(3k_{iy} - k_{sy}) + \delta(-k_{sy} - k_{iy})\right]. \quad (2.59)$$

In this case, three beams are produced. In analyzing the position of the beams it is helpful to look back at the apparatus used to produce the grating, Fig. 2.16.

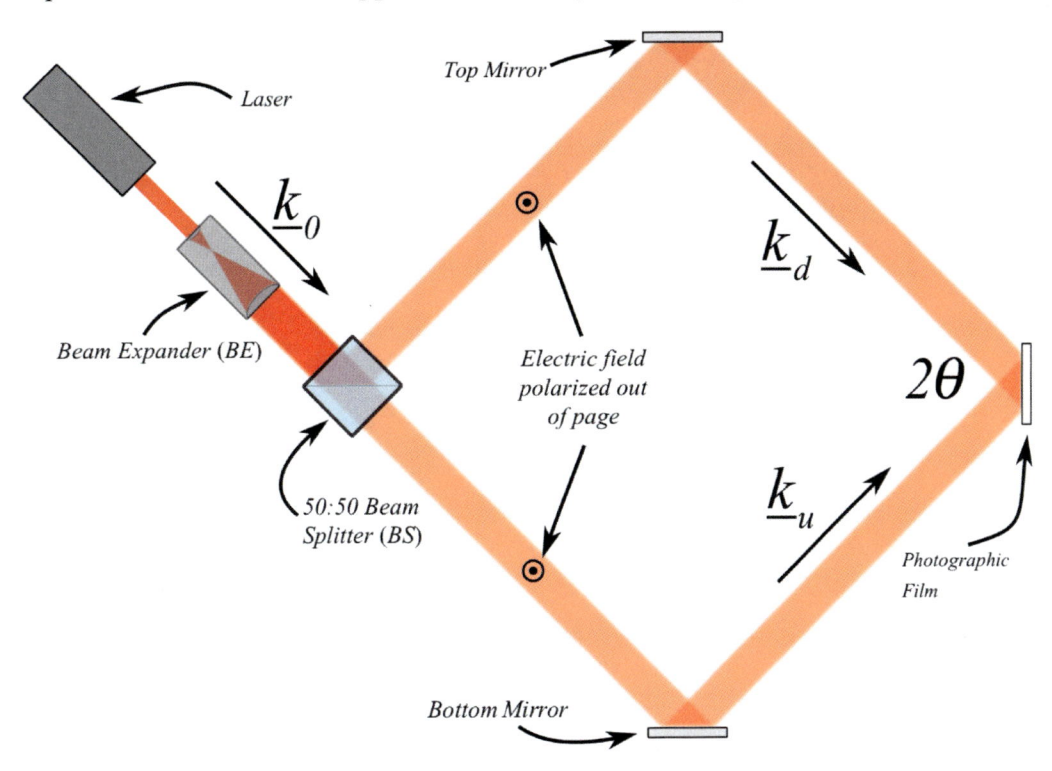

Figure 2.16 – The setup used to construct the holographic grating.

Now imagine that after we produce the grating we replace it into the position originally occupied by the photographic film, turn off the top beam (indicated by $\underline{k}_d$), and only allow the bottom beam ($\underline{k}_u$) to reach the grating. Upon passing through the grating the incident beam will be seen by an observer on the right. However, two other beams appear as seen in Fig. 2.17. One of these is a reconstruction of the beam that was turned off, while the other is of no interest.

The reconstructed beam is precisely what Holography is about. If the mirror on the top had been replaced by a toy train, then its scattering would combine with the light from the bottom mirror to form the interferogram. In this case the entire train would be reconstructed by the developed interferogram, even though the train was removed. Such an interferogram is known as a Hologram. This invention won the Nobel Prize in Physics for Dennis Gabor in 1971 (Gabor, 1971).

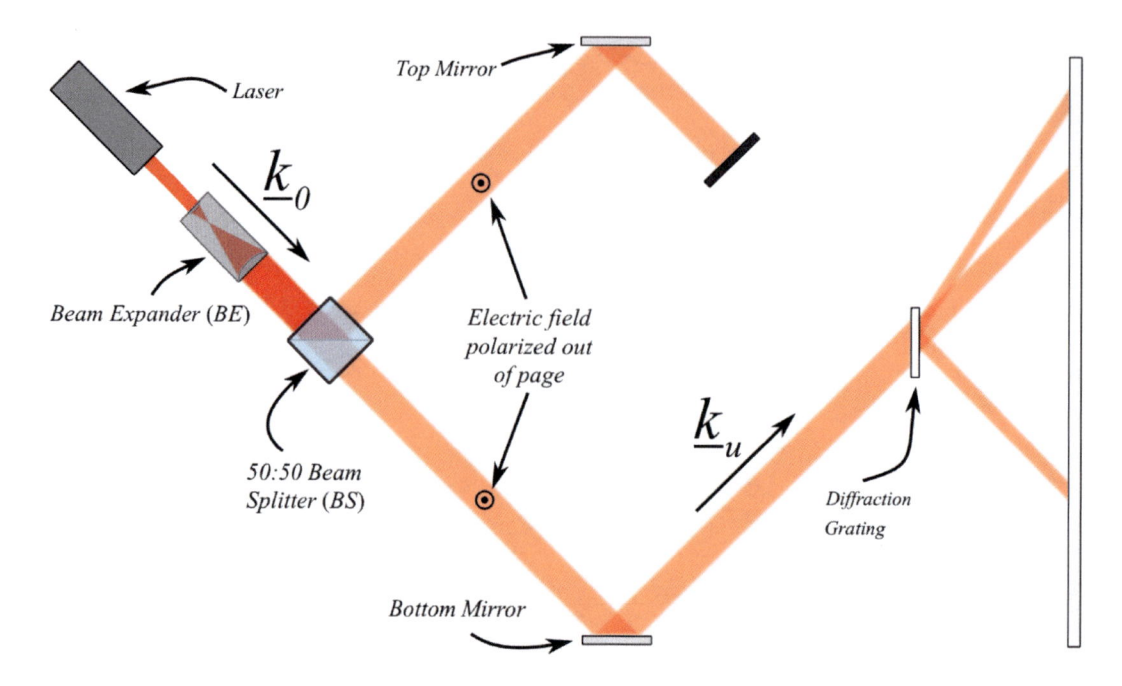

Figure 2.17 – The three beams created by light diffracting through our developed interferogram after putting it back in place, while blocking one of the beams. Note the blocked beam is reconstructed.

2.8 – Grating Equation at Oblique Incidence

In the previous discussion of oblique incidence the incident angle was equal to the angle θ that produced the grating pitch. We now ask what happens if the oblique incident light is at an arbitrary angle θ_i to a grating having an arbitrary pitch d.

The angular position of the mth order diffracted beam θ_m for a grating illuminated perpendicular to its surface is predicted by a "grating equation" that beginning physics students commit to memory, $d\sin(\theta_m) = m\lambda$, where d is the spatial pitch (i.e. $d = P_s$). When the incident light impinges obliquely on the grating at an arbitrary angle θ_i as shown in Fig. 2.18 below, the result is different and more general.

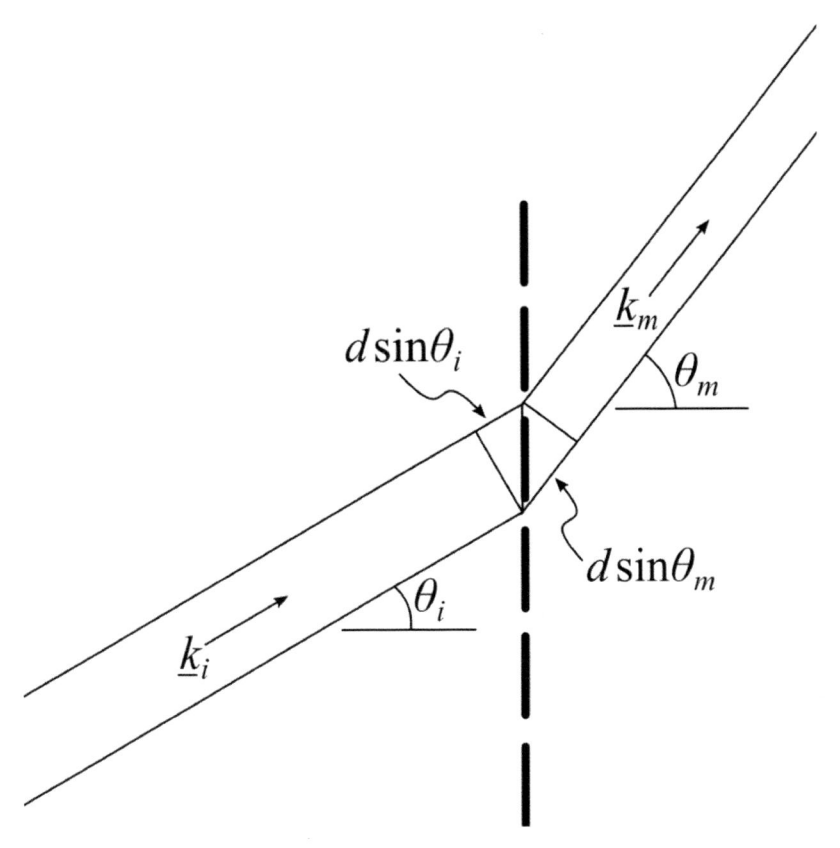

Figure 2.18 – Diffraction grating illuminated at Oblique Incidence.

What it means to have a diffracted beam of order *m* is that in going from an incident wavefront to the diffracted beam wavefront, the light travels *m* wavelengths longer through an adjacent slit. For example, in Fig 2.18 it should be clear that the path is longer through the lower slit than through the upper slit by $d[\sin(\theta_m) - \sin(\theta_i)]$. For the path to be *m* wavelengths longer requires that

$$d[\sin(\theta_m) - \sin(\theta_i)] = m\lambda. \qquad (2.60)$$

Note that Eq. (2.60) reduces to the more common grating equation at normal incidence. The order *m* can be either a positive or negative integer. At normal incidence there is a clear symmetry in as much as $\theta_{-m} = -\theta_m$, however for oblique incidence that symmetry is lost.

2.9 – X-ray Diffraction and the Structure of DNA

X-ray diffraction was developed in the 20th century and was a key factor in the rapid growth in knowledge of life at the cellular and molecular level. The method was first used to analyze the structure of crystals and is sometimes called X-ray crystallography. The structure of a crystal is periodic with repeated unit cells. The Fourier transform of these periodic structures shows points, called Bragg scattering peaks, with maximum intensity ideally proportional to the square of the number of repeating unit cell. These points are the result of constructive interference and have an intensity that is easily detected. One can determine the structure of a crystal by analyzing the spacing between Bragg peaks. Early researchers realized that it was theoretically possible to analyze a non-periodic structure using X-ray diffraction; however, because of the low symmetry diffraction spots are less discrete than for a periodic structure.

Throughout the 20th century the tools and techniques of crystallography were improved enormously. There have been 29 Nobel Prizes in physics, medicine or chemistry awarded for work related to X-ray diffraction. Perhaps the most interesting and controversial of these is the 1962 Nobel Prize in Medicine awarded to Francis Crick, James Watson, and Maurice Wilkins for identifying the helical structure of DNA. The work of Rosalind Franklin, a physical chemist and X-ray crystallographer, who skillfully imaged the diffraction pattern produced by DNA, undoubtedly aided their efforts. Franklin and her assistant Raymond Gosling developed an elaborate technique with which to hydrate DNA and pull it into strands (Franklin, 1953). The DNA strands were then mounted to a support and exposed to X-rays for around 62 hours. All the while the strands had to maintain constant humidity because this form of DNA is found to vary slightly depending on its hydration level and such a change would have blurred the diffraction image. An illustration of the apparatus is shown in Fig. 2.19.

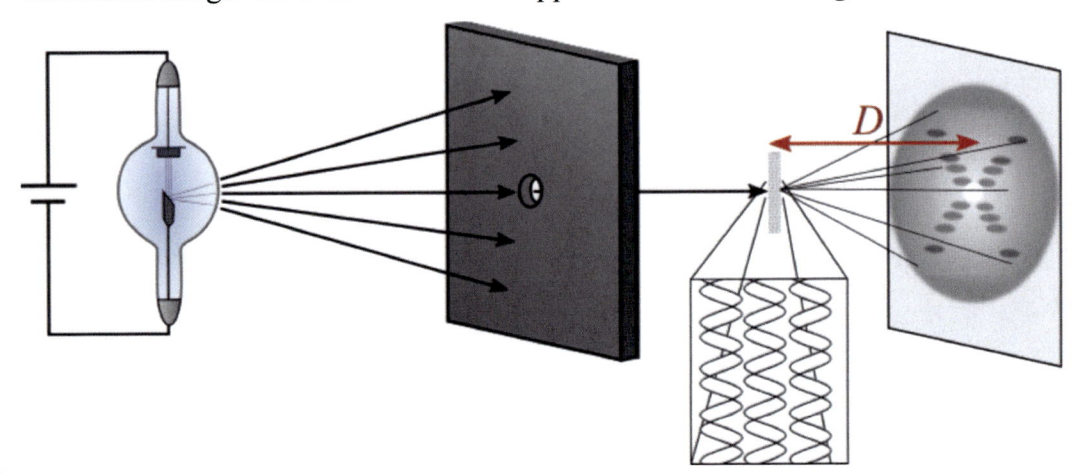

Figure 2.19 – The scattering approach used by Franklin and Gosling to produce Photo 51, which was imaged at a distance D from the DNA sample.

Overall the experimental techniques were very advanced for the time. Franklin and Gosling performed their experiment meticulously and were rewarded with a beautiful image, now famously known as Photo 51. It is shown in Fig. 2.20.

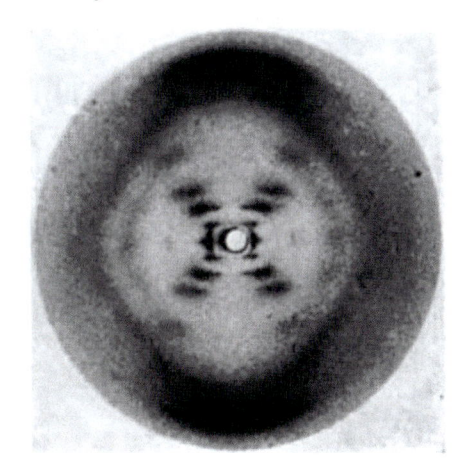

Figure 2.20 – Photo 51, the image obtained from X-ray diffraction of DNA that revealed its helical structure.

Unfortunately, Rosalind Franklin passed away at the age of 37 in 1958 and was not included in the group awarded the 1962 Nobel Prize as the honor is not awarded posthumously. The diffraction pattern shown in Photo 51 is characteristic of a double helix and knowledge of this diffraction image allowed Watson and Crick to finalize their model for DNA (Watson and Crick, 1953).

While the pattern shown in Photo 51 is quite interesting it is not obvious to the untrained eye that it is indicative of a helical structure. Let's look at some notable features of the pattern and see what we can make of them. The original image has been marked up for this purpose as seen in Fig. 2.21.

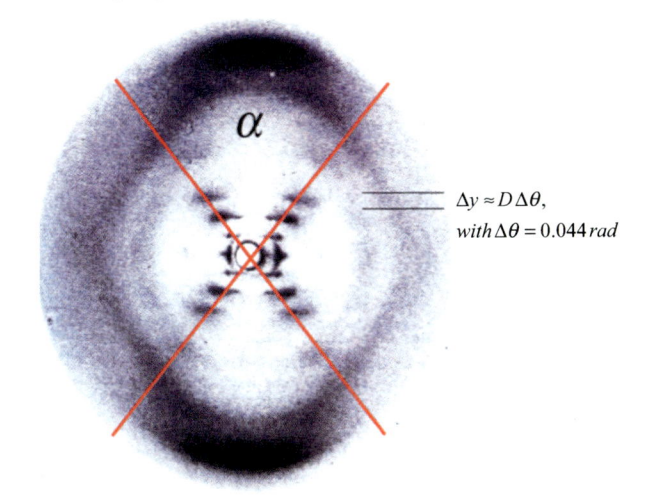

$\Delta y \approx D \Delta \theta,$
$\text{with } \Delta \theta = 0.044 \, rad$

*Figure 2.21 – A markup of Photo 51, where **α** is the angle formed by the cross.*

There are three points of interest that are present in the image:

-The X-shape with angle $\alpha = 0.99$ rad.

-The scattering angle between layer lines of $2.5° = 0.044$ rad.

-The missing 4th maxima (layer line).

These three features make it possible to decode the structure of DNA. We will now do some detective work to determine the structure of DNA from the image.

The first thing to notice in Fig. 2.21 is the cross that is formed by drawing lines through the spotted regions of high intensity; the image is a negative, regions of high intensity are black. This cross is the signature of a helical structure; this fact was familiar to Crick, and is due to scattering from tangents of a 2D projection of the phosphate backbone, as seen in Fig. 2.22.

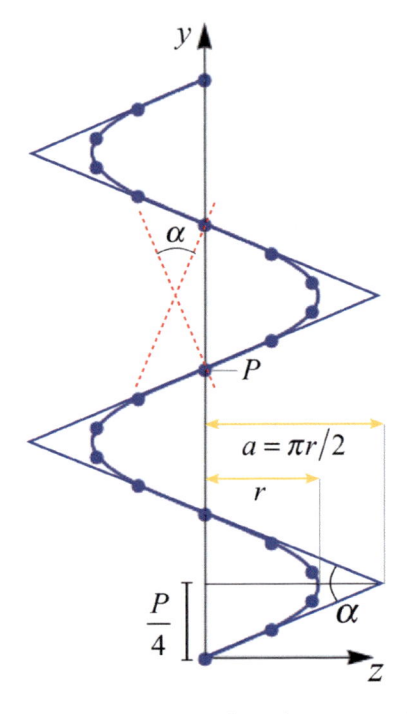

Figure 2.22 – Aperture treatment of DNA strand with tangent lines added to the sine wave. The aperture you used for homework actually provides a closer match to Photo 51, but the geometry of this aperture allows us to make handy measurements related to the pitch of the helix.

One might wonder why it is possible to think in terms of a planar aperture such as Fig. 2.22 when discussing scattering from a helical structure. The reason lies in examining Eq. (2.44), which is reproduced here for convenience:

$$c(\underline{q}) = \int_{Vol} A(\underline{r})e^{-i\underline{q}\cdot\underline{r}}d^3\underline{r}. \qquad (2.61)$$

For simplicity we may consider each of the atoms that scatter as delta functions; $A(\underline{r}) = \sum_i f(\underline{r}_i)\delta(\underline{r}-\underline{r}_i)$. On this basis the integral is discretized as

$$c(\underline{q}) = \sum_i f(\underline{r}_i)e^{-i\underline{q}\cdot\underline{r}_i}, \qquad (2.62)$$

and the intensity at a particular q is $|c(q)|^2$. The reason why we can consider a projection of the atoms onto a plane perpendicular to the incident wave vector (yz-plane), has to do with the direction of the differential scattering vector $\underline{q}$ for small scattering angles. For this circumstance $\underline{q}$ is approximately parallel to the y-z plane (Fig. 2.22), so that a position vector component of an atom along the incident direction (x-direction), will add minimally to the phase in Eq. (2.62); the phase is of the form $\underline{q}\cdot\underline{r}_i$, so only the y and z components of the atom's position are important.

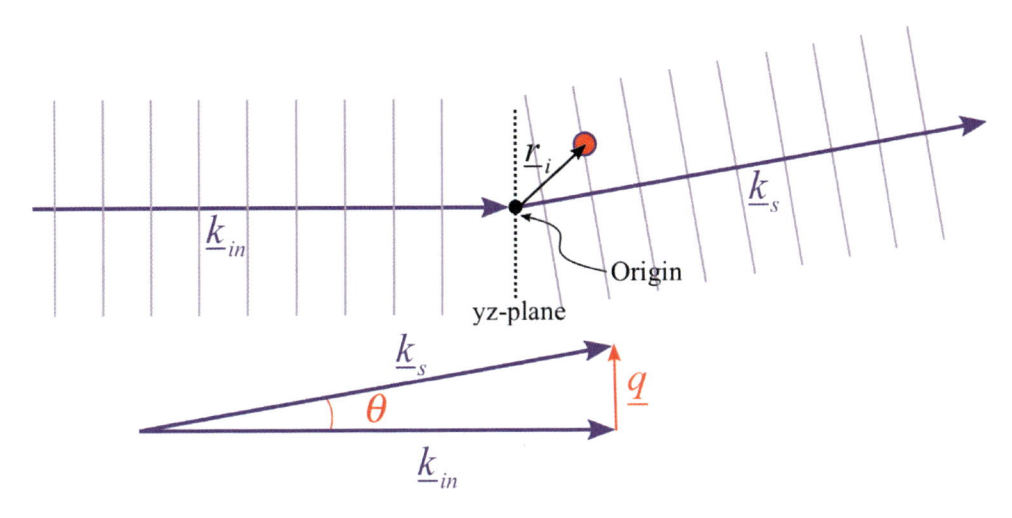

Figure 2.23 – For low angle scattering, $\underline{q}$ is nearly perpendicular to the incident wave vector. As a result, components of an atom's position in the direction of incident wave vector make a neglible contribution to the phase in Eq. (2.62), i.e. to $\underline{q}\cdot\underline{r}_i$.

Going forward we will follow this prescription in attempting to simulate Photo 51 (Exercise 2.10).

An additional observation in Photo 51 is that the light circle in the center is a hole that was made in the photographic film at the point where the radiation was most intense. Without the hole the high intensity of the unscattered X-ray beam would have bleached the film and made it much more difficult to view the off axis scattering. Horizontal lines drawn through the regions of high intensity are called layer lines. The X-rays used in the experiment are from Cu K$-\alpha$ and have wavelength $\lambda = 0.15\ nm$. The angle formed by the cross is $\alpha = 0.99\ rad$, as indicated in Fig. 2.21. By assuming that the layer lines correspond to the period, or pitch, of the helix, P, as if it were a grating, we find $P\sin(\theta) = m\lambda$. Since the separation in radians is small, the angle between interference orders, $\Delta\theta$, is λ/P, from which P can be estimated;

$$P \approx \frac{\lambda}{\Delta\theta} = \frac{0.15\ nm}{0.044\ rad} = 3.4\ nm. \qquad (2.63)$$

We can use the angle α formed by the cross to determine the radius r of the helix. As an exercise, use Fig. 2.21 to show that

$$\frac{r}{P} = \frac{\cot(\alpha/2)}{2\pi}. \qquad (2.64)$$

The angle α has been measured and the pitch has been estimated in Eq. (2.63), these two results can now be used to estimate the radius of the helix;

$$r = \frac{P\cot(\alpha/2)}{2\pi} = \frac{3.4\ nm \cdot \cot(0.99/2)}{2\pi} = 1\ nm. \qquad (2.65)$$

The final clue that we must analyze is the absence of a 4th order maximum in the images. In Photo 51 three layer lines are clearly present, with the one closest to the center being rather weak. However, where the fourth layer line may be expected to produce a region of high intensity none is present. This is indicative of a phase shift between two gratings for which the aperture function of one of the helices is $A_1(y)$. Let's represent the transform of the first grating as

$$c_1(k_y) = \int A_1(y) e^{-ik_y y}\ dy. \qquad (2.65)$$

Because we assume that the two DNA strands are identical the pattern produced by the second grating is the same as that produced by the first but with an added phase shift of $\phi = -k_y fP$, where f is a fraction of the helical pitch P;

$$c_2(k_y) = c_1(k_y) e^{-i\phi}. \qquad (2.66)$$

The total transform produced by the double helix is then

$$c_T(k_y) = c_1(k_y) + c_2(k_y) = c_1(k_y)\left[1 + e^{-i\phi}\right]. \tag{2.67}$$

This can be rewritten as

$$c_T(k_y) = 2e^{-i\phi/2}\cos(\phi/2)c_1(k_y). \tag{2.68}$$

We are interested in determining the phase difference ϕ that eliminates the 4th-order maximum in the diffraction pattern. From this we will be able to determine the spatial displacement between the two helices. To start we take the square of the absolute magnitude of Eq. (2.69) ;

$$\left|c_T(k_y)\right|^2 = 4\cos^2(\phi/2)\left|c_1(k_y)\right|^2. \tag{2.69}$$

We see that normal diffraction from one grating is modulated by our phase shift.

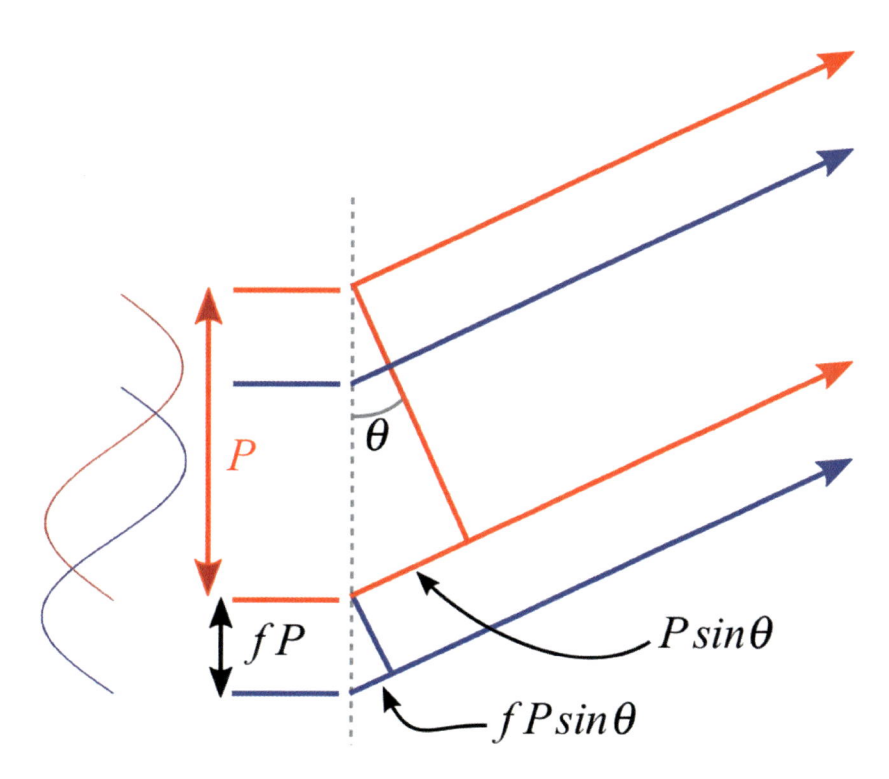

Figure 2.24 – The separation between the strands produces phase differences in the diffracted light.

In Fig. 2.24 the red and blue slits represent the two helices, each with period P_s. These slits are displaced by some distance relative to each other that can be represented as a fraction of the period, $f \cdot P_s$. The diffraction maxima produced by two slits on a distant screen was discussed in Section 2.4. The positions on these maxima are related to the phase differences in the scattered light. If we think of a single strand of DNA, or just the red slits in Fig. 2.23, the phase difference is

73

$k_0 P_s \sin(\theta)$ and maxima occur where $k_0 P_s \sin(\theta) = 2\pi m$. These maxima correspond to the intensity peaks along the layer lines in Fig. 2.20. For a single strand of DNA there will be an intensity maximum at $m = 4$, but inspecting Photo 51 clearly shows this is not the case for the double strand. This is because of the phase shift between the strands, i.e. the $\cos^2(\phi/2)$ term in Eq. (2.69). As a reminder the phase shift

$$\phi = -k_0 f P_s \sin(\theta). \qquad (2.70)$$

For the 4$^\text{th}$-maximum $k_0 P_s \sin(\theta) = 4 \cdot 2\pi$ so

$$\phi = 8\pi f. \qquad (2.71)$$

To produce the necessary zero we require that

$$\cos^2\left(\frac{\phi}{2}\right) = \cos^2\left(\frac{8\pi f}{2}\right) = 0. \qquad (2.72)$$

This condition is met when $f = 1/8$ or $f = 3/8$, as you should check for yourself. A useful way to visualize this is by plotting $\cos^2(\phi/2)$ and $|c_T(k_y)|^2$ separately, as seen in Fig. 2.24.

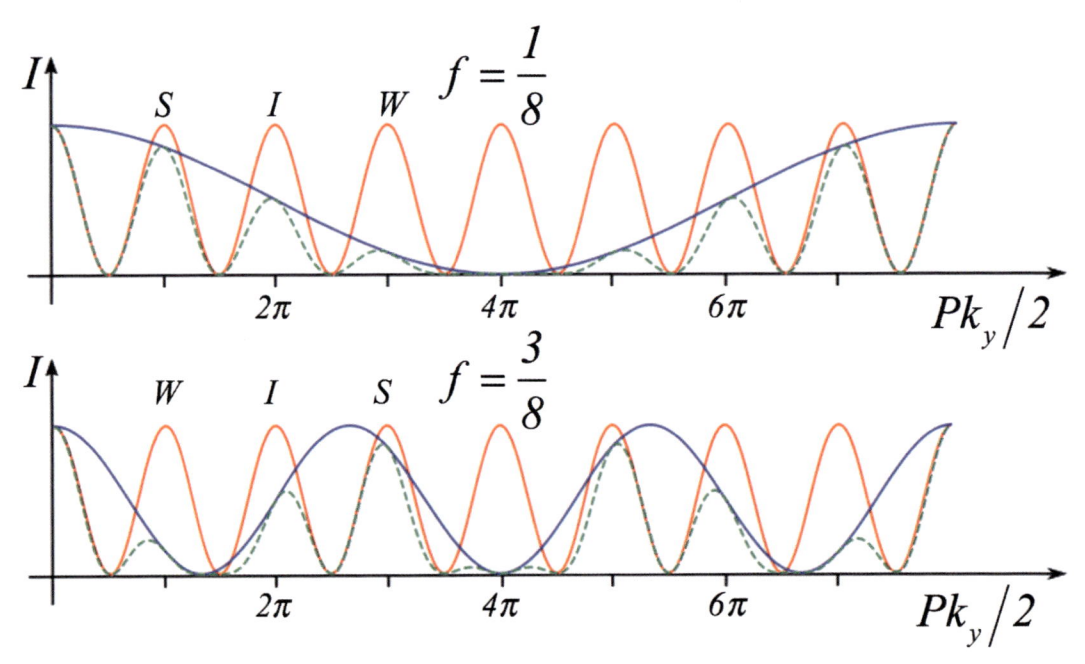

Figure 2.24 – The intensity patterns produced by a single strand of DNA (red) and contribution from the relative phase shift between strands (blue). The green curve shows the product of the red and blue curves. This product for $f = 3/8$ closely resembles the relative intensities of the different orders in Photo 51; marked S, I, and W for strong, intermediate, and weak, respectively.

In Fig. 2.24 the shifts corresponding to $f = 1/8$ and $f = 3/8$ clearly cancel the 4th-order maximum. However, by closely comparing Photo 51 with theory we find that the weak 1st-order maximum in Photo 51 fits $f = 3/8$ better.

2.10 – Chapter Two Exercises

Exercise 2.1 – Suppose you would like to form a grating with a pitch of 1 micron using a laser with a wavelength $\lambda = 650 nm$.

a) What angle θ in Fig. 2.16 and in Eq. (2.17) will be needed ?

b) Using Eq. (2.60) find the diffracted angle θ_{-1} for the $m = -1$ beam. How does its magnitude compare with θ ?

c) What are θ_0 and θ_{+1}.

d) Aim you 650 nm laser at the 1000 line per mm grating at angle θ to confirm the positions of θ_{-1}, θ_0 and θ_{+1}. For best results you may need to mount your grating on the black plug that is provided with the kit, and insert the plug into the bearing. Angular measurement may be easier this way, since the base board has printed angles on it.

Exercise 2.2 – Imagine you send light through an aperture that transmits as a Gaussian with maximum transmission at its center;

$$A(y) = e^{-|a|y^2}.$$

Use the Fourier Transform to find the diffracted field produced by this aperture. If you would like, you may use Wolfram Alpha. Do you note anything interesting about the function and its transform?

Exercise 2.3 – *Double Slit Aperture.* Suppose a laser having wavelength of 1 μm is shined at normal incidence onto the aperture $A(y)$ as shown in Fig. 2.26. The figure displays two slits each 10 μm in width and having their centers separated by 20 μm.

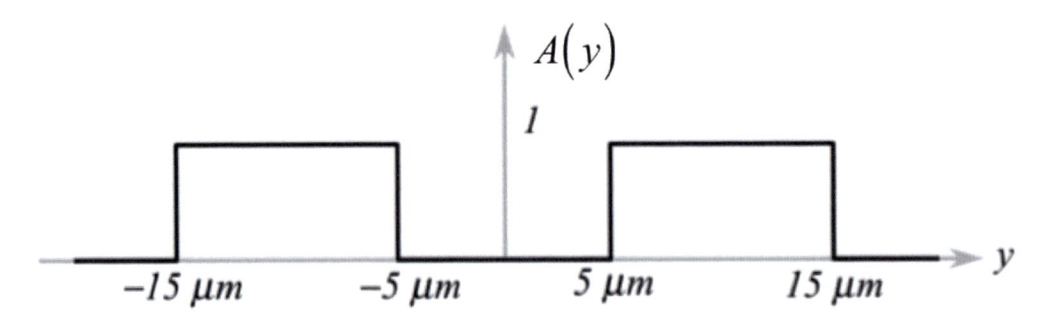

Figure 2.26 – A double slit aperture.

Compute the Fourier transform $c(k_y)$ of this aperture function from Eq. (2.25) and plot the intensity $|c(k_y)|^2$ pattern produced by this aperture.

Exercise 2.4 – A typical laser has beam width $L = 2$ mm, and most gratings have ~ 1000 lines per mm, therefore a laser beam incident on the grating will cover ~ 2000 lines. The aperture function for the grating we discussed in Section 2.2 is

$$A(y) = \frac{1}{2} + \frac{1}{2}\cos(k_g y),$$

where $k_g = 2\pi/\lambda_g$, with λ_g being the grating pitch ($\lambda_g = 2\mu m$ in this case). A section of the aperture function is shown in Fig. 2.27.

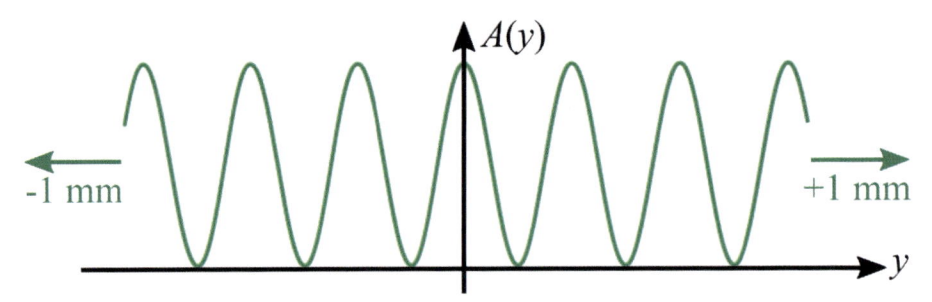

Figure 2.27 – The aperture function for a diffraction grating.

Calculate the diffracted intensity, and plot it.

Exercise 2.6 – The lowest order mode of a cylindrical laser beam such as that which produces the focused skirt in Fig. 1.20 has a planer wave-front at the center of the focal region which acts as a cylintrical aperture function,

$$A(y,z) = e^{-(y^2+z^2)/w_0^2} = e^{-\rho^2/w_0^2},$$

where w_0 is the cylindrical displacement at which the field drops by 1/e of its value at the origin. By treating x as the axial coordinate starting from the center of the focal region, describe how the beam spreads as x increases. Hint- You will need to take a Fourier transform of this aperture function.

Exercise 2.7 – Fig. 2.28 shows a simple model of the position of carbon atoms in Benzene. Given this geometry, you would like to calculate the scattering pattern from this Benzene model. The hydrogen atoms have been excluded in the figure since scattering is proportional to the square of the number of electrons in an atom; the atomic number squared, Z^2. So hydrogen has roughly 1/36 the scattering power of Carbon. Write an aperture function for this Benzene model using 2D Delta functions to describe the individual atoms. For this problem let $a = 0.14$ nm.

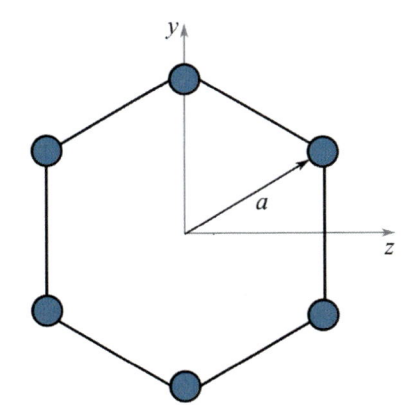

Figure 2.28 – Model of a benzene ring.

Exercise 2.8 – Take your aperture function from Ex. 2.7 and calculate the diffracted intensity pattern from an X-ray source with wave vector $k\hat{x}$. Make a diffraction image of this pattern with abscissa k_z and ordinate k_y.

Exercise 2.9 – Find the diffracted image generated by the red strand of the DNA (shown below) after modeling the aperture function as a sum of 20 two dimensional delta functions. Length dimensions are in nanometers (nm) with the vertical spacing between base pairs in the red to blue strand being 0.34nm, and the radius of the helix equal to 1nm.

$A = 20$ $R = 1 nm$

Red Strand: $m = 1,...,A$ $\theta_m = \dfrac{4\pi m}{A}$

$y_m = m \cdot 0.34$ $z_m = R\cos(\theta_m)$

Blue Strand: $m = 21,...,2A$ $\theta_m = \dfrac{4\pi(m-A)}{A}$

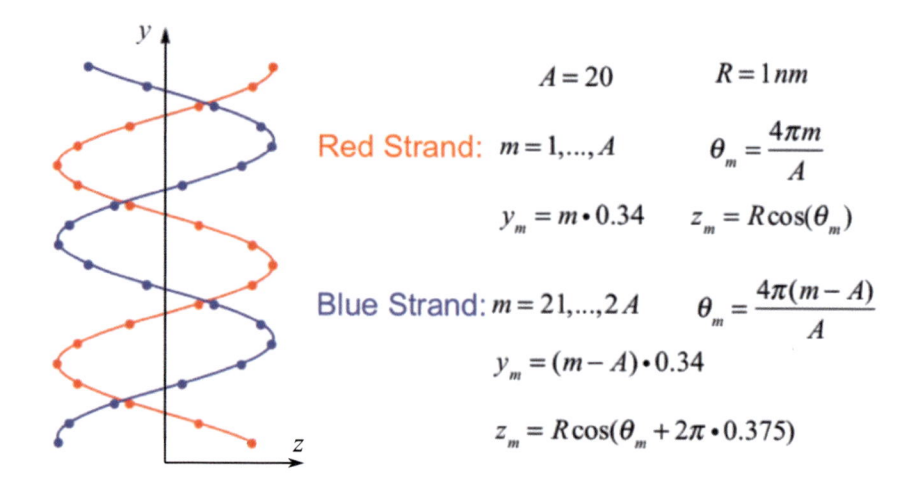

$y_m = (m-A) \cdot 0.34$

$z_m = R\cos(\theta_m + 2\pi \cdot 0.375)$

Exercise 2.10 – DNA is composed of many more atoms than included in the aperture modeled above. Compare the diffraction pattern generated in Exercise 2.9 with Photo 51, Fig. 2.20. Are many of the same features present? If so, how does our simplified model composed of delta functions capture some of the same features as the far more complex structure?

Exercise 2.11 – Try to analyze the diffraction pattern produced by both strands (red and blue) by taking $f = 3/8$, as in Fig. 2.24. Is this pattern a better match to Photo 51? See (Lucas, 1999).

Exercise 2.12 – Derive Eq. (2.64) for the ratio of the radius r to the pitch P of dsDNA.

Exercise 2.13 – Rosalind Franklin's Photo 51 is shown in Fig. 2.19. The image is blackened where X-rays have fallen. You will note that the 4th diffraction spot from the center does not appear. From this Watson and Crick concluded that there are two helices displaced from each other by 3/8 of a helical pitch P. Given the same experimental setup, what displacement would they have concluded if the 2nd spot were missing instead of the 4th?

Exercise 2.14 – DNA has a pitch of 3.4 nm and in Rosalind Franklin's experiment scattered light the x-ray wavelength was 0.15 nm. How far from the DNA sample would a screen need to be placed to be in the Fraunhofer region?

Experiment #2 – Spectroscopy: laser's wavelength using diffraction gratings.

Measure the wavelength of your Laser using the two transmission diffraction gratings in the Pocket Optics kit. The number of lines per millimeter of each of the gratings are written on the grating. You should be able to arrive at the wavelength using Eq. (2.39).

With the Pocket Optics Baseboard, my setup is shown below. Note that the Grating is inserted into slot G, and the Metal 6 inch ruler is inserted into slot E. The distance between the grating and the ruler along the centerline is 100mm. A diffraction spot on the ruler (not shown), allows one to to determine the angle of diffraction.

Experiment#2 Setup

Scan for a video demo of this experiment

79

Experiment #3 – Metrology: determine pitch of CD and DVD when used as reflection gratings.

Experimentally determine the line spacing for a CD and DVD. Treat each as a reflection grating. Use the laser wavelength λ determined from Experiment #1. Direct your laser perpendicular (normal) to the CD or DVD surface and measure the angle θ of the first order back-scattered intensity peak relative to this normal. The spacing between the tracks d follows the familiar equation $d\sin\theta = \lambda$.

With the Pocket Optics baseboard, my setup is shown below. The diffraction spot occurs on the side of the ruler facing the DVD disc (not shown). The distance between the DVD disc and the ruler in the photo is 50mm.

Scan for a video demo of this experiment

Experiment#3 Setup

Chapter Three – The Classical Atom and Dielectric Theory

3.1 – Snell's Law, Origin of Refractive Index, Dispersion, and Chromatic Aberration

When dealing with diffraction we learned how light scatters from different objects. In doing this we represented the objects as apertures that are described in terms of the opaqueness or transparency. In this chapter we would like to refine our study of light-matter interactions to describe how light interacts with various materials. At a material interface monochromatic light is bent according to Snell's law as seen in Fig. 3.1. For example, the light entering glass from vacuum bends closer to the normal in accordance with $\sin(\theta_2) = \sin(\theta_1)/n$, where n is the refractive index of the material. The refractive index for glass at visible wavelengths of greater than one, is in accordance with the observation that θ_2 is less than θ_1. If instead of having light incident from vacuum we had used a material such as water, a more general form for Snell's law would be

$$n_1 \sin(\theta_1) = n_2 \sin(\theta_2), \qquad (3.1)$$

where n_1 and n_2 are the refractive indices associated with the incident and refracting mediums, respectively. Snell arrived at Eq. (3.1) by experiment. Later in this text, we will derive it theoretically using Maxwell's equations. In what follows, we will examine some of the important properties of refractive index.

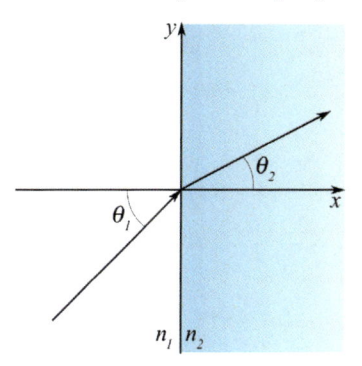

Figure 3.1 – A ray of light travelling in a medium with refractive index n_1 is bent inward towards the normal when entering a material of higher refractive index, n_2. Here we have ignored reflection from the interface. We will elaborate on reflection in Chapter 5.

You may know that light is refracted differently depending on its frequency/wavelength. One consequence is the effect that a prism has on white light (Fig. 3.2). As discovered by Newton, white light is broken into a rainbow of colors by a glass prism as a consequence of the varying refractive index of the glass with wavelength.

Figure 3.2 – White light can be decomposed into different frequencies, or colors, using a prism. This has to do with how the refractive index of a medium relates to wavelength of the light travelling through it.

The change in refractive index of light with wavelength, known as dispersion, has been measured carefully for many types of glass (Fig. 3.3). Notice that the refractive index of all of the glasses in this figure decrease as the wavelength is increased from the near ultraviolet to the infrared.

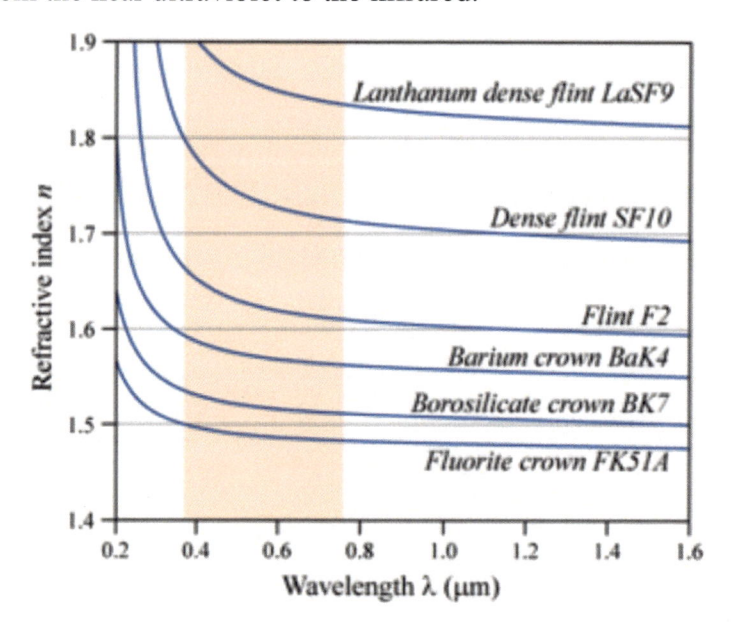

Figure 3.3 – The refractive index of many materials depends on the wavelength of light (Image from Wikipedia-DrBob-Licensed under Creative Commons Attribution-Share Alike 3.0 via Wikimedia Commons).

Dispersion is important when dealing with optical instruments. For example, the focal length of a lens may depend on wavelength. The focal length of the converging lens shown in Fig. 3.4 is shorter in the blue than in the red. This effect, called chromatic aberration, results in the distortion of multicolor images.

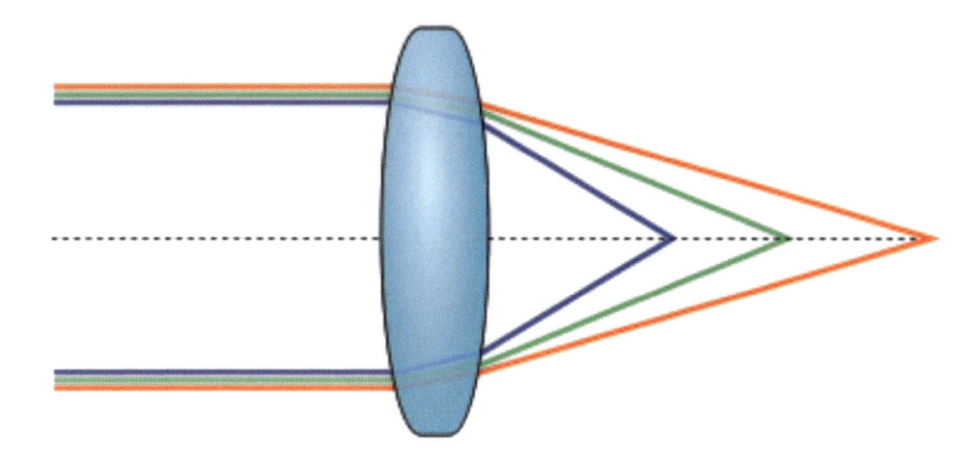

Figure 3.4 – Light at different frequencies is focused at different points by a lens, this is called chromatic aberration.

It is interesting and important to note that the refractive index in a given dielectric medium defines the speed of a travelling light wave through the medium. Simply stated, the speed of light in a given medium, v_m, is the speed of light in vacuum divide by the associated refractive index;

$$v_m = \frac{c}{n}. \qquad (3.2)$$

Refractive index is such an important subject that it begs the question, "what causes it on an atomic level?" To answer this question we will present a semi-classical description of the atom, from which the interaction of light with matter can be built. Out of this will come a physical description of refractive index and dispersion.

3.2 – The Lorentz Atom

In order to explain phenomena such as dispersion and chromatic aberration we must develop a model for the atoms with which light interacts. Without requiring a background in Quantum Mechanics a classical atom can be built. To start the discussion I must emphasize that "light walks on two feet"[11] (i.e. Electric and Magnetic), and that the light force that pushes an atom requires both fields. The

---

[11] A term often used by Martin Wegener of the Karlsruhe Institute of Technology.

classical atom, now known as the Lorentz Atom, was first introduced by Hendrik Antoon Lorentz in 1878 to extend Maxwell's theory. A depiction of this model is shown in Fig. 3.5.

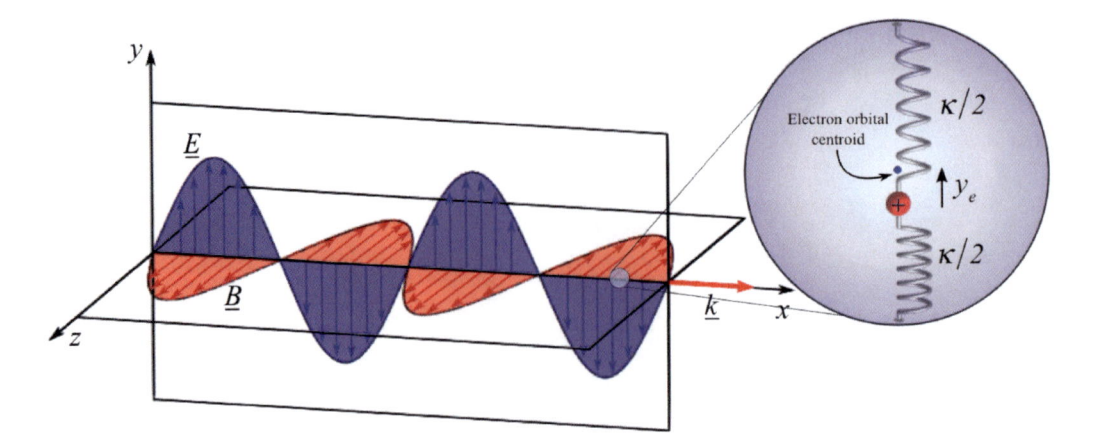

Figure 3.5 – A Lorentz atom can be modeled by imagining an electron and proton system in which the force on the electron acts as a system of two springs with equilibrium position centered at the stationary proton.

The original Lorentz atom, also known as a Lorentz oscillator, involved an electron interacting with a much larger positive nucleus through a Hooke's law potential, while having an electromagnetic wave impinging on it. In more modern terms we depict the Lorentz atom as a semi-classical structure with a tiny positive nucleus having charge $|q_e|$ connected to a dimensionally larger negatively charged orbital with charge $-|q_e|$, by Hooke's law springs. The springs with effective force constant κ are associated with the electron-nucleus interaction. Since the nucleus is much more massive than the electron, it is relatively stationary by comparison to the electron under stimulation by the light field, $E_0 \cos(\omega t)$. This allows us to concentrate on the electron's motion; on the centroid of the negative charge. This oscillatory mode is accompanied by a radiation loss that we represent in this mechanical model by a drag coefficient, c_d. Together this semi-classical model depicted in Fig. 3.5 results in a dynamical equation for the *y*-component of the electron's motion,

$$m_e \frac{d^2 y_e}{dt^2} = -\kappa\, y_e - c_d \frac{dy_e}{dt} - |q_e| E_0 \cos(\omega t). \qquad (3.3)$$

We simplify this equation by dividing by the electron mass and identifying κ/m_e with the square of the undamped harmonic oscillator frequency ω_0^2, while the damping rate c_d/m_e is identified as γ;

$$\frac{d^2 y_e}{dt^2} = -\omega_0^2 y_e - \gamma \frac{dy_e}{dt} - \frac{|q_e|}{m_e} E_0 \cos(\omega t), \qquad (3.4)$$

To get the solution to this equation long after the transient behavior disappears it is convenient to represent $\cos(\omega t)$ as $e^{-i\omega t}$, consistent with the temporal phase we normally use for a plane wave. Since Eq. (3.4) is linear, the response y_e can have no other frequency than ω. Consequently we represent y_e as $y_{e0} e^{-i\omega t}$. One consequence is that y_{e0} will now be complex, but the appropriate solution is its real part since the field on the right side of the equation is the real part of $E_0 e^{-i\omega t}$. With this representation for the displacement substituted into Eq. (3.4), we find that

$$-\omega^2 y_{e0} = -\omega_0^2 y_{e0} + i\omega\gamma y_{e0} - \frac{|q_e|}{m_e} E_0 \qquad (3.5)$$

from which we determine that

$$y_{e0} = \frac{-\dfrac{|q_e|}{m_e} E_0}{-\omega^2 + \omega_0^2 - i\omega\gamma}. \qquad (3.6)$$

The velocity of the electron is $v_e = dy_e/dt$, which is the same as $-i\omega y_e$;

$$v_{e0} = \frac{i\omega \dfrac{|q_e|}{m_e} E_0}{-\omega^2 + \omega_0^2 - i\omega\gamma}. \qquad (3.7)$$

If this were all that is happening there could be no radiation pressure on our classical atom to push it in the direction of the light (i.e. x-direction). However there is more, the Lorentz force. The oscillating electron in a magnetic field, gives rise to a Lorentz force, $\underline{F} = -|q_e|\underline{v}_e \times \underline{B}$. The velocity in Eq. (3.7) is in the y-direction, $\underline{v}_e = v_e \hat{y}$, but the magnetic field will be in a direction appropriate to the wave moving in the positive x-direction with its electric field polarized along y;

85

$\underline{B}$ is in phase with the electric field and in the z-direction. On this basis the time-averaged force on our classical atom is in the x-direction and given by

$$\langle F_x \rangle_t = -|q_e| \langle v_{ey} B_z \rangle_t. \tag{3.8}$$

Taking the time average is straight forward for a product of phasors

$$\langle F_x \rangle_t = -\frac{|q_e|}{2} \text{Re}\left[v_{ey} B_z^* \right] = -\frac{|q_e|}{2c} \text{Re}\left[v_{ey} E_y^* \right]. \tag{3.9}$$

In the last expression we have used the fact that the z-component of the magnetic field is equal to the y-component of the electric field divided by the speed of light. Now it is only a matter of combining Eq. (3.9) with Eq. (3.7). Before doing this Eq. (3.7) needs to be cleaned up a bit so that the real and imaginary components of the velocity can be separated. To do this we must multiply its numerator and denominator on the RHS of Eq. (3.7) by the complex conjugate of the denominator. In this form the velocity is

$$v_{y0} = \frac{i\omega \frac{|q_e|}{m_e} E_{y0} \left(\omega_0^2 - \omega^2 + i\omega\gamma \right)}{\left(\omega_0^2 - \omega^2 \right)^2 + \left(\omega\gamma \right)^2}. \tag{3.10}$$

Combing Eq. (3.10) and Eq. (3.9) gives

$$\langle F_x \rangle_t = -\frac{|q_e|}{2c} \text{Re}\left[\frac{i\omega \frac{|q_e|}{m_e} E_{y0} \left(\omega_0^2 - \omega^2 + i\omega\gamma \right)}{\left(\omega_0^2 - \omega^2 \right)^2 + \left(\omega\gamma \right)^2} E_{y0}^* \right]. \tag{3.11}$$

Taking the real part gives the radiation force

$$\langle F_x \rangle_t = +\frac{|q_e|^2}{2cm_e} \frac{\omega^2 \gamma E_{y0}^2}{\left(\omega_0^2 - \omega^2 \right)^2 + \left(\omega\gamma \right)^2}. \tag{3.12}$$

What we see is that we have a force in the direction of the light that is proportional to the intensity of the light. We also see that as ω approaches ω_0 the force goes through resonance. Using our standard form for intensity [Eq. (1.70)],

$$\langle F_x \rangle_t = \frac{|q_e|^2}{\varepsilon_0 m_e c^2 \gamma} \left[\frac{(\omega\gamma)^2}{\left(\omega_0^2 - \omega^2 \right)^2 + \left(\omega\gamma \right)^2} \right] \langle I \rangle_t. \tag{3.13}$$

I have balanced the frequency dependent term in brackets so that it is dimensionless. If you check the units you will see that the RHS of Eq. (3.13) has the units of a force; i.e. Newtons.

3.3 – Radiation Force and Absorption Cross Section

Recall that if absorption is the origin of the momentum exchange that leads to the radiation force,[12] the force will be equal to the power absorbed divided by the speed of light;

$$\langle F_x \rangle_t = \frac{P_a}{c}. \qquad (3.14)$$

If we represent the power absorbed by the atom as a cross section for absorption σ times the intensity $\langle I \rangle_t$ incident on the atom, then the radiation force

$$\langle F_x \rangle_t = \frac{\sigma \langle I \rangle_t}{c}. \qquad (3.15)$$

By comparing Eq. (3.15) with Eq. (3.13) we have an opportunity to make a theoretical estimate for the absorption cross-section

$$\sigma = \frac{|q_e|^2}{\varepsilon_0 m_e c \gamma} \left[\frac{(\omega \gamma)^2}{(\omega_0^2 - \omega^2)^2 + (\omega \gamma)^2} \right]. \qquad (3.16)$$

This can be rewritten as

$$\sigma = \frac{|q_e|^2}{\varepsilon_0 m_e c \gamma} \frac{1}{\left[1 + \left(\frac{(\omega_0 - \omega)(\omega_0 + \omega)}{\gamma \omega} \right)^2 \right]}. \qquad (3.17)$$

I have simply factored $(\omega_0^2 - \omega^2)/(\gamma \omega)$ in order to find a simple approximation. Absorption principally occurs near resonance, $\omega = \omega_0$, where the frequency dependent expression in Eq. (3.17) is 1. Away from resonance the denominator of this expression can get much larger than 1, depending on the relaxation rate γ. For most absorbing atoms γ is less than one millionth of ω_0 so by just

---

[12] Emission from an atom is isotropic leading to no final average radiation momentum.

"detuning" the frequency by 1%, $(\omega_0 - \omega)/\gamma$ grows to 10,000, while there is hardly any change in $(\omega_0 + \omega)/\omega$ (i.e. it remains very near 2). Consequently, the cross-section falls to 1/20,000 of what it was at resonance. Using the reasonable approximation near resonance that $(\omega_0 + \omega)/\omega \approx 2$, the frequency dependence in Eq. (3.17) can be written as

$$\frac{1}{1+\left(\frac{(\omega_0-\omega)(\omega_0+\omega)}{\gamma\omega}\right)^2} \approx \frac{1}{1+\left(\frac{(\omega_0-\omega)}{\gamma/2}\right)^2} = \frac{(\gamma/2)^2}{(\gamma/2)^2+(\omega_0-\omega)^2}. \quad (3.18)$$

The function on the far right is known as a Lorentzian line shape. Plotted in Fig. 3.6 is the Lorentzian vs. detuning frequency for an arbitrary line width γ.

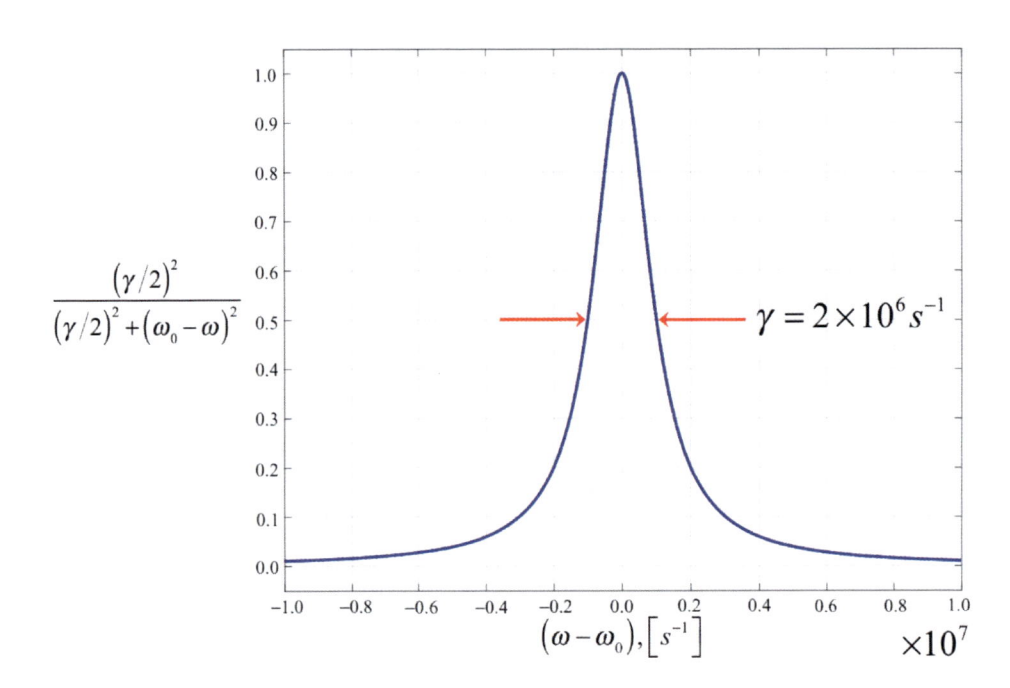

Figure 3.6 – Lorentzian shape.

In terms of the Lorentzian form in Eq. (3.18), Eq. (3.17) now gives a simple expression for the absorption cross-section,

$$\sigma = \frac{|q_e|^2}{\varepsilon_0 m_e c \gamma} \frac{(\gamma/2)^2}{(\gamma/2)^2+(\omega_0-\omega)^2}. \quad (3.19)$$

In the next few sections we will continue to develop this model of the atom. I encourage you to become familiar with it because we will rely on it throughout the rest of the text. Exercises 3.1-3.4 are a good way to begin this study.

> **Comments: Absorption Cross-Section** – Atoms and molecules are quantum mechanical (QM), so it is important to understand the limitations of the theory presented thus far. Surprisingly the frequency dependence of the cross section in Eq. (3.19) has the same form as that given by QM, and the cross section is accurate if the quantum system only has a single transition, the analog of which is the resonance of our Lorentz atom. However, the Quantum reality is that there is a finite probability for the electron to transition to a number of different "states", each having a different cross section. On this basis only a fraction of the classical cross section would be allowed to any one state. This fraction, f_{abs}, is known as the "oscillator strength" of the transition. For a dye molecule, f_{abs} is close to one.
>
> In the fourth exercise at the end of this chapter (Ex. 3.4) you will be asked to estimate the maximum cross-section for our Lorentz atom for a value of $\gamma = 2 \times 10^{14} s^{-1}$. Your answer will be near $5.3 \times 10^{-20} m^2$, which is on the order of the physical area of a typical dye molecule, however atomic cross sections can be much larger since relaxation rates can be much smaller. So long as the oscillator strength is near one, the maximum cross section depends principally on the relaxation rate γ.

3.4 – Lorentz Atom Spring Constant and Semi-Classical Atom Size

In Section 3.1 we discussed the Lorentz model for a classical atom. Here we would like to address that model in greater detail. We might start by asking, what is the origin of the spring constant in the Lorentz model? Well, it is associated with the attraction of the electron back to the nucleus. But why does this attraction simply grow linearly with electron displacement? After all Coulomb's law is nonlinear (i.e. the force would be expected to get weaker as y increases). A simple way to understand this disparity is to assume the electron charge is distributed within a sphere; sort of like the 1s orbital, from quantum mechanics. The truth is that the field distorts this spherical shape, however in what follows we will show that the simple assumption of a spherical shape does generate a linear restoring force.

As you likely recall from chemistry, the electron actually occupies an orbital; it has a probability of occupying a particular region of space. For example, the unperturbed state of hydrogen has the electron existing as a probability density within a spherical ball of radius a. In effect, an external electric wave polarized in the y-direction shakes the ball of negative charge up and down periodically. To figure out the force of attraction between the cloud of charge and the nucleus, we displace the centroid of negative charge from the nucleus (Fig. 3.7).

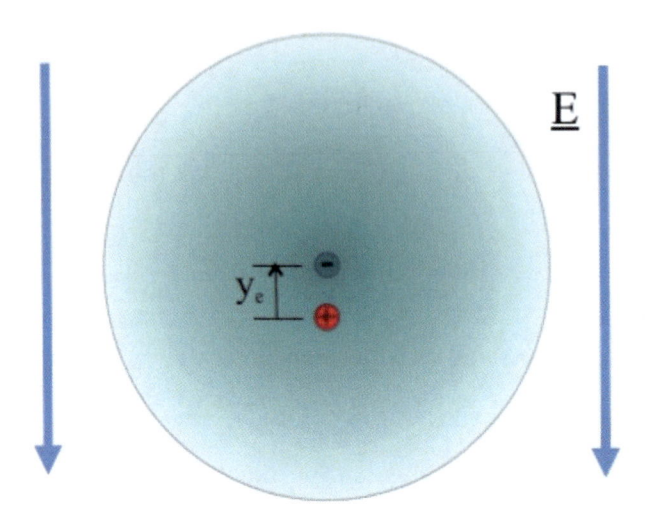

Figure 3.7 – A snapshot of the displacement between the nucleus (red dot) and centroid of negative charge at a moment when the electric field is down, and the driving frequency is well below resonance.

We suppose that the negative globe of charge has a uniform charge density $-|\rho_e|$, so that the force, which pulls the negative charge back toward the positive nucleus, can be obtained handily from Gauss's law,

$$F_e = -|q_e|\frac{|\rho_e|y}{3\varepsilon_0}.$$ (3.20)

A major feature of this force is that it takes a Hooke's law form, $F_e = -k_{LA} y_e$, with the Lorentz atom spring constant

$$k_{LA} = \frac{|q_e||\rho_e|}{3\varepsilon_0}.$$ (3.21)

To calculate the oscillator frequency it is essential that we recognize that the proton is much more massive than the electron, so that it essentially remains fixed; the spring in effect pulls around the much lighter electron. With this consideration our simple classical atom oscillates with squared angular frequency $\omega_0^2 = k_{LA}/m_e$, or

$$\omega_0^2 = \frac{|q_e||\rho_e|}{3\varepsilon_0 m_e}. \tag{3.22}$$

This can be simplified further by recognizing that the volume integral of the electron density is just the charge of the electron, from which

$$\omega_0^2 = \frac{|q_e|^2}{4\pi\varepsilon_0 m_e a^3}. \tag{3.23}$$

The radius a of the semi-classical atom is

$$a = \left(\frac{|q_e|^2}{16\pi^3 \varepsilon_0 m_e f_0^2}\right)^{1/3}, \tag{3.24}$$

where f_0 is the resonant frequency.

The above is an idealized model. In truth the electron density arrived at from quantum mechanics is not uniform but falls off exponentially, and the negative cloud will not remain spherically symmetric under the distortion of the electric field. Surprisingly, the semi-classical radius will turn out to be only a factor of two off from what quantum mechanics gives (Ex. 3.6).

3.5 – Dielectric Theory, Refractive Index, and Chromatic Aberration

Dielectrics are non-conducting materials. To understand their properties it is best to look at where you likely were introduced to dielectrics in the first place. It was in the theory of capacitors. Fig. 3.8 shows a capacitor to which a battery is applied. The field in the dielectric is not only due to charges on the capacitor plates in the form of a charge density σ_c but also due to induced surface charge densities at the top and bottom of the dielectric slab. This induced charge, known as the polarization charge density σ_{pc}, is the result of the polarization of the atoms and molecules in the dielectric. It appears at the edges because of a cancellation of the charge pairs within the slab. Both the primary charge on the plates and the polarization charge determine the electric field within the dielectric slab.

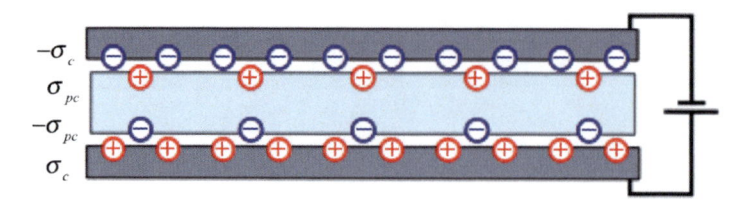

Figure 3.8 – A capacitor with a dielectric insert.

One can analyze the field in the dielectric slab region by removing the slab while imagining that the polarization charges remain in place, and treating these charges as if they were real.

The next step is to apply Gauss's law, Eq. (1.1). The net effect, which you should show as an exercise (Exercise 3.7) is that the field within the region of the dielectric slab E is

$$E = \frac{\sigma_c - \sigma_{pc}}{\varepsilon_0} . \tag{3.25}$$

where ε_0 is the so-called permittivity of free space (Note: σ_c here means charge density not cross-section). The first thing to notice is that for a given field strength, the charge on the plates is increased as the polarization charge increases

$$\sigma_c = \varepsilon_0 E + \sigma_{pc} . \tag{3.26}$$

This increases the capacitance. Often this is expressed by writing $\sigma_c = k\varepsilon_0 E$ where k is the so-called dielectric constant, which for a capacitor is greater than 1. It is common to replace $k\varepsilon_0$ with the permittivity of the material ε. On that basis

$$\varepsilon E = \varepsilon_0 E + \sigma_{pc} . \tag{3.27}$$

The interpretation of the polarization charge involves the positive charges on the dipoles terminating at the surface at the top of the slab, as shown in Fig. 3.8;

$$\sigma_{pc} = \frac{1}{A} \sum_j q_j , \tag{3.28}$$

where A is the area of the capacitor plates. By simply multiplying and dividing the right hand side of Eq. (3.28) by the dipole length δ we get the sum of dipole strength per unit volume (dipole moment $\underline{\mu} = q\underline{\delta}$);

$$\sigma_{pc} = \frac{1}{\delta \cdot A}\sum_j q_j \delta = \frac{1}{\delta \cdot A}\sum_j \mu_j. \qquad (3.29)$$

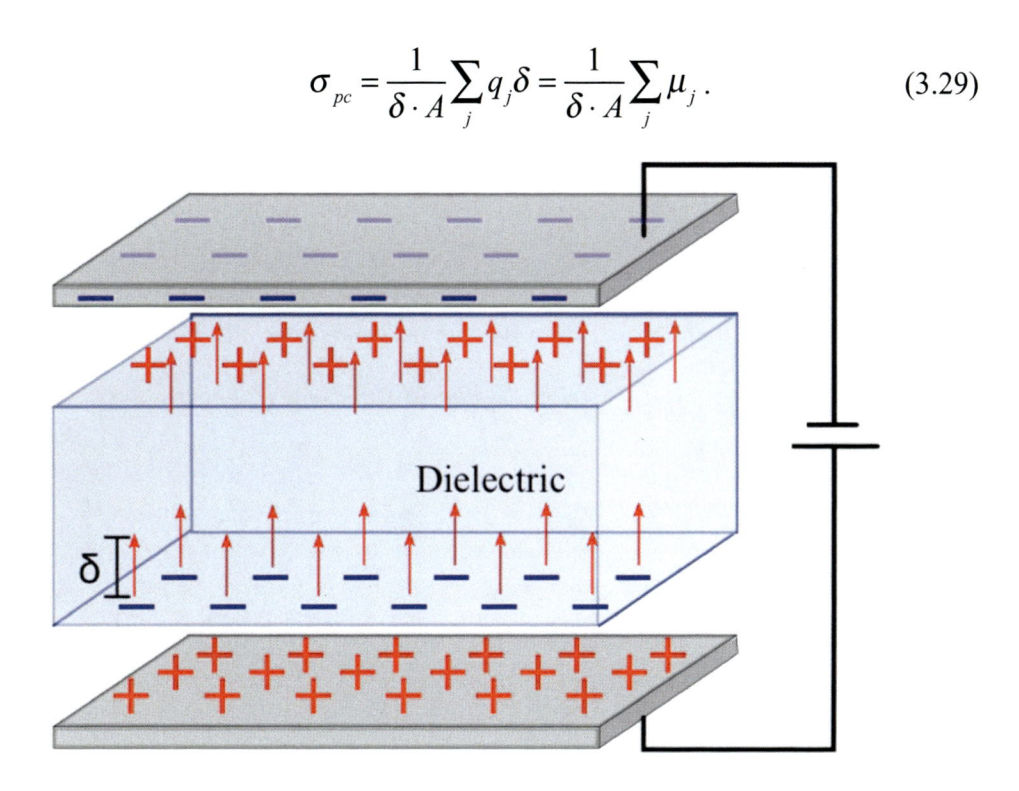

Figure 3.9 – Dielectric material with an induced polarization charge at its surface.

Fig. 3.9 shows the basic idea. When the dielectric is in the capacitor the electric field induces dipoles in this region. Because the dipoles have length δ we can consider the volume they occupy rather than the induced surface charge density. This volume density of dipole moments is known as the polarization density $\underline{P}$, a vector in the direction of the dipole moment sum;

$$\underline{P} = \frac{1}{\delta \cdot A}\sum_j \underline{\mu}_j, \qquad (3.30)$$

allowing Eq. (3.27) to be written as

$$\varepsilon \underline{E} = \varepsilon_0 \underline{E} + \underline{P}. \qquad (3.31)$$

For an isotropic material $\underline{P}$ is parallel to $\underline{E}$ so we will just consider their magnitudes,

$$\varepsilon = \varepsilon_0 + \frac{P}{E} \qquad (3.32)$$

If we applied an AC source in place of our DC battery and increased the frequency sufficiently, waves would be generated in the dielectric as the oscillating dipoles began to radiate as transmitting antennae. The velocity (v) of these waves in the dielectric is different than the velocity of light in vacuum $(c \approx 3\times 10^8 \ m/s)$, with the difference being related to the relative permittivity of the medium in relation to vacuum. This is seen in a 1D case by rewriting the wave equation, Eq. (1.43), as

$$\frac{\partial^2 E_y}{\partial x^2} = \varepsilon \mu_0 \frac{\partial^2 E_y}{\partial t^2}. \qquad (3.33)$$

Notice that ε has replaced ε_0, so the speed of light in a dielectric,

$$v = \frac{1}{\sqrt{\varepsilon \mu_0}}, \qquad (3.34)$$

while the speed of light in vacuum is

$$c = \frac{1}{\sqrt{\varepsilon_0 \mu_0}}. \qquad (3.35)$$

Combining Eq. (3.34) and Eq. (3.35) allows us to express v in terms of c;

$$v = \frac{c}{\sqrt{\varepsilon/\varepsilon_0}}. \qquad (3.36)$$

You may recall that v is usually expressed in terms of c divided by the refractive index;

$$v = \frac{c}{n}. \qquad (3.37)$$

Thus, the denominator in Eq. (3.36) is the so-called "refractive index",

$$n = \sqrt{\frac{\varepsilon}{\varepsilon_0}}, \qquad (3.38)$$

where $\varepsilon/\varepsilon_0$ is known as the "relative permittivity". Combining Eq. (3.38) with Eq. (3.32) connects the refractive index to polarization density,

$$n^2 = \frac{\varepsilon}{\varepsilon_0} = 1 + \frac{P_0}{\varepsilon_0 E_0}, \qquad (3.39)$$

where P_0 is the amplitude of the polarization density. The refractive index is elevated above the vacuum level for light in materials with $P_0/(\varepsilon_0 E_0) > 0$. Air, for example has a refractive index of ~ 1.0003, while most glasses have refractive indices close to 1.5 in the visible portion of the spectrum. Perhaps more important is that as one goes from the red to blue portion of the visible spectrum the refractive index of glass rises. This plays a critical role in optics since refractive index controls the bending of light as light travels from air to glass, which means that in the blue simple lenses have shorter focal lengths than in the red. This, as you will recall, is called chromatic aberration, as illustrated in Fig. 3.4. To understand this effect requires fashioning a material from many atoms. To create a Lorentz atom material requires that we rewrite Eq. (3.4) in terms of dipole moments. One must think of a fixed positive charge at the center of the cage (e.g. nucleus equivalent) in Fig. 3.5. Then if the electron moves down the atom develops an upward dipole moment. On this basis the dipole moment $\mu = -|q_e|y_e$; Eq. (3.4) is converted to a dynamical equation for a dipole moment by multiplying each term by $-|q_e|$,

$$\frac{d^2\mu}{dt^2} + \gamma \frac{d\mu}{dt} + \omega_0^2 \mu = +\frac{|q_e|^2}{m_e} E_0 e^{-i\omega t}. \tag{3.40}$$

Once again as a linear equation the response of the dipole will be at frequency ω, which motivates expressing μ as $\mu_0 e^{-i\omega t}$, for which the complex amplitude satisfies

$$\mu_0 = \frac{\frac{|q_e|^2}{m_e} E_0}{\omega_0^2 - \omega^2 - i\omega\gamma}. \tag{3.41}$$

With a material composed of many such dipoles at the volume number density ρ_N the amplitude of the polarization density is

$$P_0 = \frac{\rho_N \frac{|q_e|^2}{m_e} E_0}{\omega_0^2 - \omega^2 - i\omega\gamma}, \tag{3.42}$$

and the refractive index from Eq. (3.39) is

$$n^2 = \frac{\varepsilon}{\varepsilon_0} = 1 + \frac{\rho_N \frac{|q_e|^2}{m_e \varepsilon_0}}{\omega_0^2 - \omega^2 - i\omega\gamma}. \tag{3.43}$$

Eq. (3.43) applies to a dilute sample. Atomic dipoles in a solid or liquid will be influenced by other atomic dipoles around them, since the atoms are in close proximity. Correcting for this is not difficult, but for now it is more important to understand the physical reason for a complex refractive index.

The complex nature of Eq. (3.43) has to do with the fact that the polarization density $\underline{P}$ does not oscillate in phase with the field driving it due to damping and inertia. This affects a light wave travelling through the matter by changing the magnitude of the propagation wave vector $\underline{k}$. Since as refractive index increases the velocity slows, the wavelength shrinks to λ_0/n, and as consequence the magnitude of the wave vector increases as

$$k = \frac{2\pi}{\lambda_0/n} = nk_0, \qquad (3.44)$$

where λ_0 is the wavelength in vacuum. For a complex refractive index, $n = n_r + in_i$, k becomes complex. If the imaginary part of n is positive, then the wave will decay exponentially as it propagates forward. To see this, write out the form the wave must take:

$$\underline{E} = E_0 \hat{y} e^{i(nk_0 x - \omega t)} = E_0 \hat{y} e^{-n_i k_0 x} e^{i(n_r k_0 x - \omega t)}. \qquad (3.45)$$

As to the way in which the real and imaginary parts of the refractive index depend on frequency, that is a matter of solving Eq. (3.43). Since Eq. (3.43) is already for a dilute system such as a gas, the second term on the right will be much less than 1, and an approximation is in order; $(1+z)^{1/2} \approx \pm(1+z/2)$ (where both the real and imaginary parts of z are $\ll 1$). On this basis

$$n = n_r + in_i = \sqrt{1 + \frac{\frac{|q_e|^2}{\varepsilon_0 m_e} \rho_N}{\omega_0^2 - \omega^2 - i\omega\gamma}} \approx 1 + \frac{\frac{|q_e|^2}{2\varepsilon_0 m_e} \rho_N}{\omega_0^2 - \omega^2 - i\omega\gamma}, \qquad (3.46)$$

where we have discarded the unphysical negative solution. By multiplying the top and bottom of the 2nd term on the right by the conjugate of the denominator

$$n = n_r + in_i \approx 1 + \frac{\frac{|q_e|^2 \rho_N}{2\varepsilon_0 m_e}(\omega_0^2 - \omega^2 + i\omega\gamma)}{(\omega_0^2 - \omega^2)^2 + (\omega\gamma)^2}, \qquad (3.47)$$

from which we find

$$n_r = 1 + \frac{\frac{|q_e|^2 \rho_N}{2\varepsilon_0 m_e}(\omega_0^2 - \omega^2)}{(\omega_0^2 - \omega^2)^2 + (\omega\gamma)^2} \quad \text{and} \tag{3.48}$$

$$n_i = \frac{\left(|q_e|^2 \rho_N / 2\varepsilon_0 m_e\right)\omega\gamma}{(\omega_0^2 - \omega^2)^2 + (\omega\gamma)^2}. \tag{3.49}$$

These two equations are plotted in Fig. 3.10 to show their distinctive shapes. Notice that the real and imaginary parts are plotted on different axes.

Note that n_i has a Lorentzian-like form, and although n_i has the dissipation rate γ in the numerator, this is not the case for the real part. The imaginary part of the refractive index controls the spatial decay of a wave as it enters a medium, and this spatial decay is faster for larger γ. On the other hand the real part of the refractive index n_r controls the bending of the light (refraction) by glass and other substances. Resonance for most "clear" glass occurs in the ultraviolet.

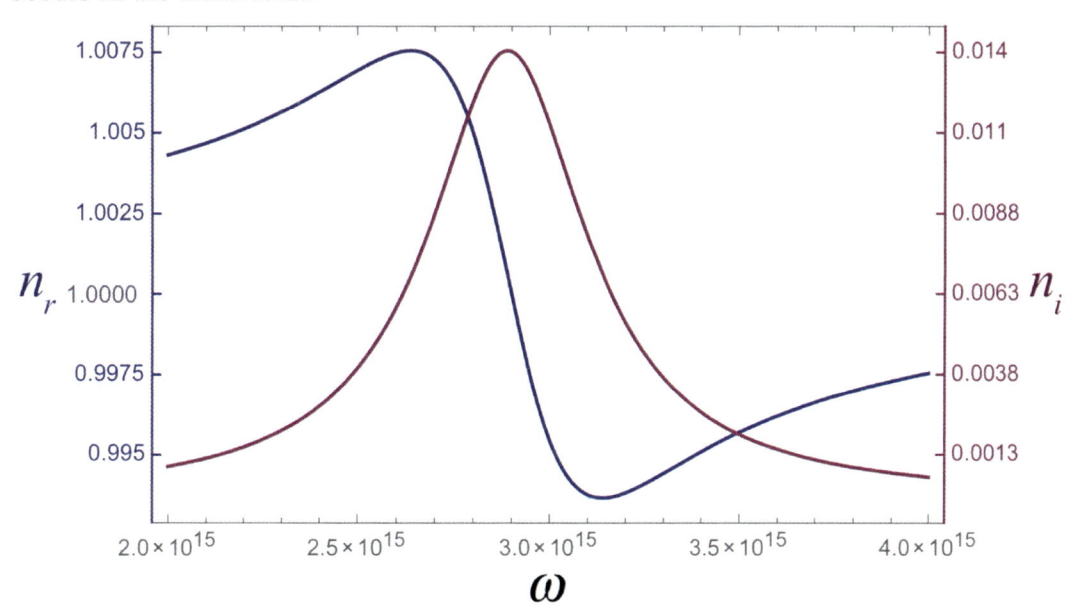

Figure 3.10 – The real and imaginary parts of the refractive index plotted with respect to frequency and centered at resonance with $\omega_0 = 2.9 \times 10^{15}\ \text{sec}^{-1}$ and $\gamma = 5 \times 10^{14}\ \text{sec}^{-1}$. The density of ions was taken to be $\rho_N = 2.4 \times 10^{25}\ m^{-3}$, approximately the density of Na atoms in a street lamp.

Our model predicts that the real part of the refractive index should increase as the blue portion of the spectrum is approached. Fig. 3.11 shows this effect for a

97

number of common glass materials. While Fig. 3.10 and Fig. 3.11 are both plots of the refractive index they take on very different shapes. This can be understood by recognizing that the refractive index is plotted against frequency in Fig. 3.10 and against the free space wavelength in Fig. 3.11.

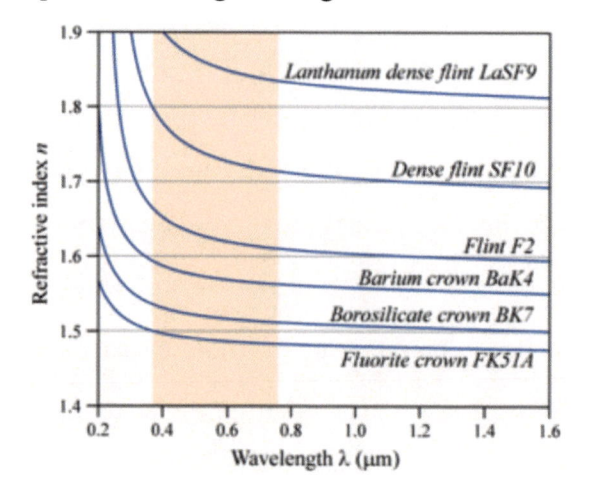

Figure 3.11 – The refractive index of many materials depends on the wavelength of light traveling through it (from Wikipedia-DrBob-Licensed under Creative Commons Attribution-Share Alike 3.0 via Wikimedia Commons).

As a consequence of the dispersion in glass the bending of light will be more pronounced for blue light than for red light for a typical glass as shown in Fig. 3.12 (recall I said that we would stick with gases for the dilute approximation, however phenomenologically the same sort of thing occurs for the more exact solution).

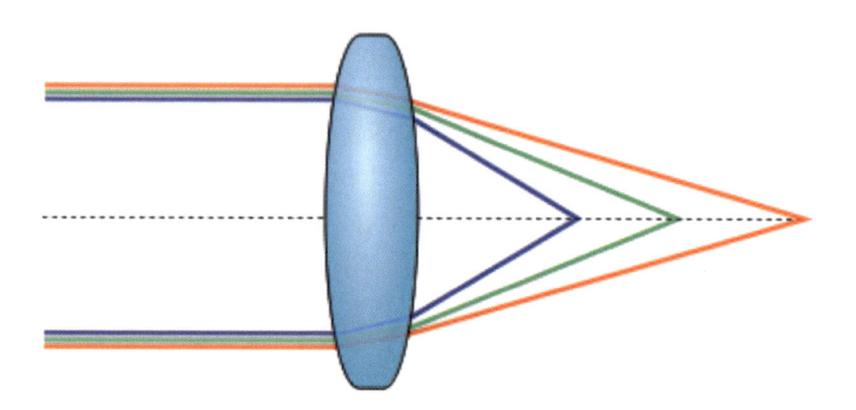

Figure 3.12 – Chromatic aberration.

3.6 – Phase and Group Velocity

Another important application of dispersion is the effect that it has on the superposition of waves in such a medium. If waves with different temporal frequencies superpose in a medium other than vacuum, the phase velocity of each will depend on its frequency, since the refractive index depends on frequency. Unlike the vacuum where all waves move at the same speed, in the dispersive medium, one wave can run ahead of the other. The superposition of these waves leads to the generation of an additional velocity, known as group velocity. It is this velocity that accounts for the speed with which information moves through the medium. Although the phase velocity can exceed the speed of light when the refractive index dips below 1 as in Fig. 3.10, the group velocity will not. It is the group velocity that ultimately counts in fulfilling Einsten postulate.

To understand the distinction between phase velocity v_p and group velocity v_g we superpose two waves

$$E(x,t) = E_0 e^{i(k_1 x - \omega_1 t)} + E_0 e^{i(k_2 x - \omega_2 t)} \qquad (3.50)$$

having different nearby temporal frequencies ω_1 and ω_2, as well as different nearby spatial frequencies k_1 and k_2. By nearby we mean that the difference in each of these quantities divided by their average is very small; $\Delta\omega/\omega_{AV} \ll 1$, where $\Delta\omega = \omega_2 - \omega_1$, and $\omega_{AV} = (\omega_2 + \omega_1)/2$. The complex vector sum for Eq. (3.50) is shown in Fig. 3.13.

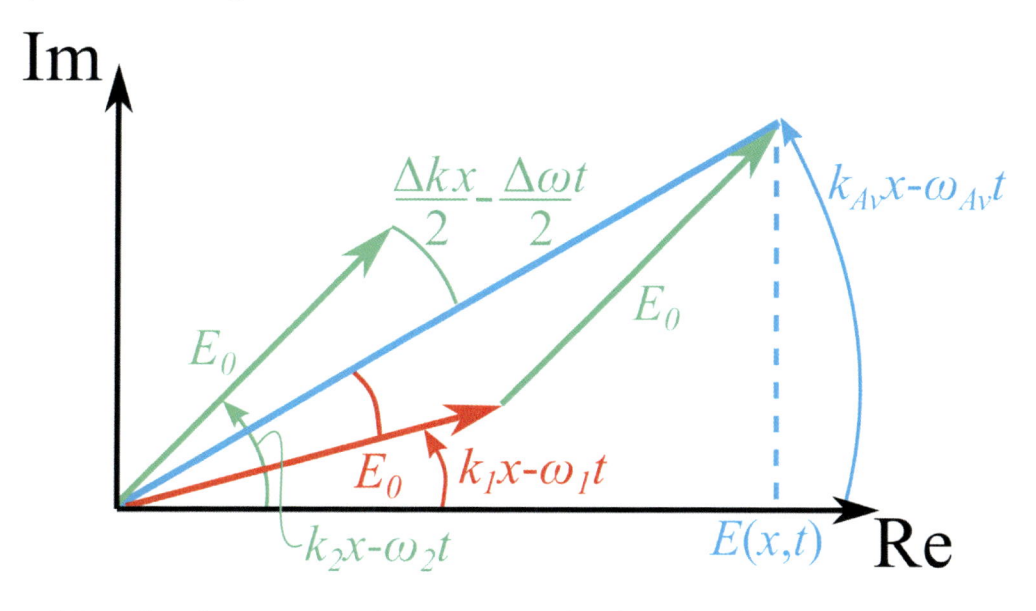

Figure 3.13 – Complex vector sum for 2 waves moving through a dispersive medium.

By using superposition in the complex plane (Sec. 2.2) the resultant length is revealed using trigonometry, with the result being $2E_0 \cos[(\Delta k/2)x - (\Delta\omega/2)t]$, where $\Delta k = k_2 - k_1$. After projecting this resultant on the real axis, the superposed field is revealed

$$E(x,t) = 2E_0 \cos[(\Delta k/2)x - (\Delta\omega/2)t] \cos(k_{AV}x - \omega_{AV}t) \;, \quad (3.51)$$

where $k_{AV} = (k_2 + k_1)/2$. Since Δk and $\Delta\omega$ are much smaller than k_{AV} and ω_{AV}, the field in Eq. (3.51) represents a long wavelength travelling wave $\cos[(\Delta k/2)x - (\Delta\omega/2)t]$ modulating a much shorter wavelength travelling wave $\cos(k_{AV}x - \omega_{AV}t)$. The latter wave travels with a velocity similar to either of the individual waves; almost the same phase velocity of either one. The much longer wavelength modulation wave, $\cos[(\Delta k/2)x - (\Delta\omega/2)t]$, travels at a velocity known as the group velocity, $v_g = \Delta\omega/\Delta k$. With the waves differentially close in frequency, the group velocity becomes $v_g = d\omega/dk$. Fig. 3.14(a) shows the waves at a fixed time. You will note that they have different wavelengths. The superposition in Fig. 3.14(b) shows a primitive "wave packet". This wave packet moves at the group velocity, whereas the wave within the envelope moves at the average phase velocity.

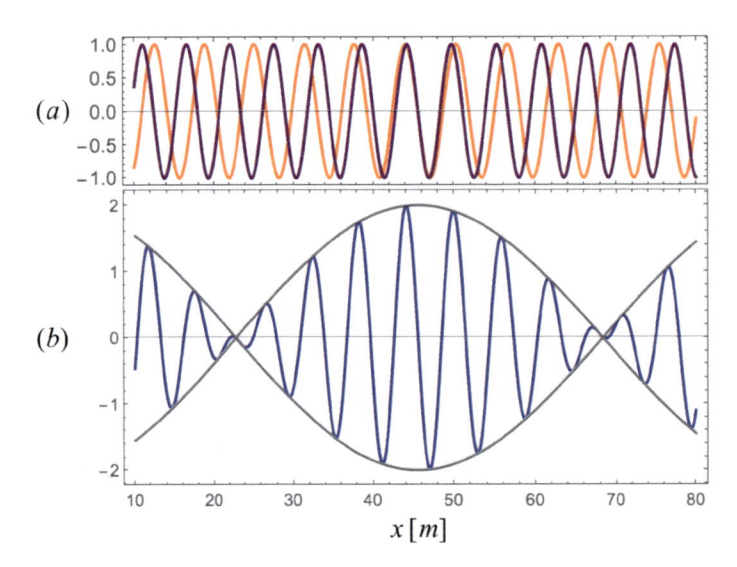

Fig. 3.14 – Superposition of waves having different frequencies at $t = 0 \, s$. The basic information required to generate the figure are the frequencies $\omega_1 = 2.90 \times 10^{15} \, s$, $\omega_2 = 3.45 \times 10^{15} \, s^{-1}$, and the refractive indices $n_1 = 1.45$, $n_2 = 1.47$.

Information can be transmitted by modulation of a carrier wave. That can involve the superposition of additional frequencies. In glass, within the visible spectrum, the refractive index increases with frequency. That means that the information will be transmitted at the group velocity. To compute the group velocity in relation to the phase velocity, we write $\omega = kc/n(\omega)$, and then compute $v_g = d\omega/dk$. We will leave this as an exercise, however its result

$$v_g = d\omega/dk = v_p - (kc/n^2) dn/dk = v_p + (\lambda c/n^2) dn/d\lambda, \quad (3.52)$$

shows that if the refractive index increases (decreases) with frequency (wavelength), as it does in a typical glass, the group velocity will be smaller than the phase velocity.

Comments: Complex Refractive Index – Now that we are at a point of applying the complex refractive index it is important to take some time to discuss how Quantum Mechanics (QM) changes Eq. (3.43). It is important to note as before (see Comments: Absorption Cross Section, Sect. 3.2) that the basic frequency dependence of a given transition is not changed, however an electron in a QM atom has many states to which it can be excited, with each characterized by a different transition frequency ω_m, damping rate γ_m, and oscillator strength F_m. The overall effect is that Eq. (3.43) now has the form

$$n^2 = \frac{\varepsilon}{\varepsilon_0} = 1 + \frac{\rho_N |q_e|^2}{\varepsilon_0 m_e} \sum_m \frac{F_m}{\omega_m^2 - \omega^2 - i\omega\gamma_m}, \quad (3.53)$$

where

$$\sum_m F_m = 1. \quad (3.54)$$

So long as only a single transition exists, Eq. (3.53) reduces to Eq. (3.43).

3.7 – Beer-Lambert Law: light attenuation in solution

It is apparent that the imaginary part of the refractive index n_i in Eq. (3.49) bears some resemblance to the optical cross-section σ in terms of the frequency dependence. Just as a reminder I will write σ as it would have been without the Lorentzian simplification,

$$\sigma = \frac{|q_e|^2}{\varepsilon_0 m_e c \gamma} \left[\frac{(\omega \gamma)^2}{(\omega_0^2 - \omega^2)^2 + (\omega \gamma)^2} \right]. \quad (3.55)$$

By comparing σ with n_i in Eq. (3.49) we see that the two are proportional to each other, more specifically

$$n_i = \left[\frac{\rho_N c}{2\omega} \right] \sigma = \left[\frac{\rho_N}{2 k_0} \right] \sigma. \quad (3.56)$$

This explains why n_i leads to attenuation of the intensity in Eq. (3.45). If we think about the attenuation as caused by each absorbing atom/molecule taking a bite out of the incoming intensity of light, then the power absorbed in each interaction is $\sigma \cdot I$, which leads to reduced intensity after the interaction. Since intensity is proportional to $E^* E$, light entering a cuvette filled with a number concentration ρ_N of absorbing molecules (Fig. 3.13) will have its intensity reduced as it traverses its light path d in the cuvette.

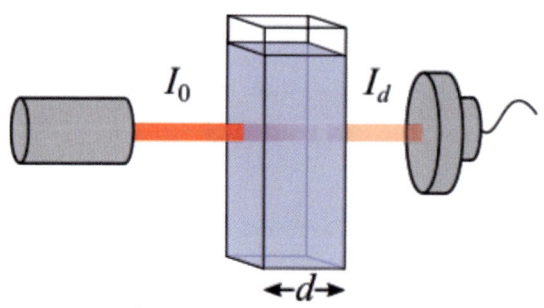

Figure 3.13 – Light attenuation through a solution in a cuvette.

This view of the attenution of light is known as the Beer-Lambert Law or simply Beer's Law. Using Eq. (3.56) in Eq. (3.45) we see a revealing exponent,

$$\frac{I_d}{I_0} = \frac{(E^* \cdot E)_d}{(E^* \cdot E)_0} = \exp(-\rho_N \sigma d). \quad (3.57)$$

This exponent $\rho_N \sigma d$ is known as absorbance, and can be extracted from a measurement by taking the natural log of I_0 / I_d. Beer's Law is commonly applied to absorbing molecules suspended in solvent (e.g. aqueous or organic).

Accurately determining the absorbance requires that attenuation associated with reflection from the cuvette windows be accounted for. This is done by dividing each transmitted intensity by the transmitted intensity through a cuvette filled with the solvent; I_0 in Eq. (3.57) becomes the intensity of light transmitted through the solvent (e.g. pure water) rather than the incident intensity. Once having measured the intensity ratio, all you have to know to get the absorption cross section is the concentration of molecules in solution. This is often given in terms of molarity (moles/liter). The number density ρ_N can be obtained by converting each mole to the appropriate number of molecules, and each liter to volume in cm^3. Of course to make the measurement you will have to build a simple instrument. If you have a Pocket Optics kit, much of the construction is already done; see Experiment #3 at the end of this chapter, where you will be asked to verify Beer's law by measuring the absorption through cuvettes having several concentrations of copper sulphate. With the data acquired in the experiment you will also be able to determine the absorption cross section of the Cu2+ ion at the frequency of your laser.

3.8 – Chapter Three Exercises

Exercise 3.1 – Show that the ratio of the magnitude of the magnetic force, $-|q_e|\underline{v}_e \times \underline{B}$, to the electric force, $-|q_e|\underline{E}$, for the light-atom interaction illustrated in Fig. 3.5 is v_e/c.

Exercise 3.2 – Show that the integral under the absorption cross-section in Eq. (3.19),

$$\sigma = \frac{|q_e|^2}{\varepsilon_0 m_e c \gamma} \frac{(\gamma/2)^2}{(\gamma/2)^2 + (\omega_0 - \omega)^2},$$

from a detuning, $(\omega_0 - \omega)$, of $-\infty$ to $+\infty$ is independent of the line width of the resonance.

Exercise 3.3 – Calculate a numerical value for the integral arrived at in Ex. 3.2 [Hint: This only depends on the properties of the electron and the speed of light].

Exercise 3.4 – Calculate the maximum cross-section from Eq. (3.16). [Hint: This does depend on the line width γ. Assume that it γ is $2\times 10^{14} s^{-1}$ (typical for a dye in solution)].

Exercise 3.5 – Show that Eq. (3.20),

$$F_e = -|q_e|\frac{|p_e|y}{3\varepsilon_0},$$

can be arrived at from Gauss's Law.

Exercise 3.6 – Calculate the size of the semi-classical atom using Eq. (3.24),

$$a = \left(\frac{|q_e|^2}{16\pi^3 \varepsilon_0 m_e f_0^2}\right)^{1/3},$$

by taking f_0 to be the lowest frequency of light that can excite hydrogen from its ground state ($1s \to 2p$, $f_0 = 2.5\times 10^{15} Hz$). You may be surprised to find that your answer agrees within a factor of 2 of the accepted value for the radius of the ground state of hydrogen.

Exercise 3.7 – Show that the field within a dielectric as given by Eq. (3.25),

$$E = \frac{\sigma_c - \sigma_{pc}}{\varepsilon_0},$$

can be obtained from Gauss's Law.

Exercise 3.8 – Starting with Eq. (3.43),

$$n^2 = \frac{\varepsilon}{\varepsilon_0} = 1 + \frac{\rho_N \frac{|q_e|^2}{m_e \varepsilon_0}}{\omega_0^2 - \omega^2 - i\omega\gamma},$$

show that the refractive index for a frequency well below resonance (i.e. wavelength well above resonance) may be approximated by

$$n^2 \approx 1 + \frac{A\lambda^2}{\lambda^2 - \lambda_0^2}. \tag{3.58}$$

Hint: Since $\omega_0^2 - \omega^2 \gg \omega\gamma$, you can ignore $\omega\gamma$ in Eq. (3.43). This effectively eliminates an imaginary part for n.

Exercise 3.9 – Find the exact solutions for n_r and n_i starting with Eq. (3.43),

$$n^2 = \frac{\varepsilon}{\varepsilon_0} = 1 + \frac{\rho_N \frac{|q_e|^2}{m_e \varepsilon_0}}{\omega_0^2 - \omega^2 - i\omega\gamma},$$

and show that in our approximate dilute limit these solutions return to Eq. (3.48),

$$n_r = 1 + \frac{\frac{|q_e|^2 \rho_N}{2\varepsilon_0 m_e}(\omega_0^2 - \omega^2)}{(\omega_0^2 - \omega^2)^2 + (\omega\gamma)^2},$$

and Eq. (3.49),

$$n_i = \frac{\frac{|q_e|^2 \rho_N}{2\varepsilon_0 m_e}\omega\gamma}{(\omega_0^2 - \omega^2)^2 + (\omega\gamma)^2}.$$

Exercise 3.10 – You would like to know the absorption cross section of the organic dye Cy5 and 650 nm. Fortunately you can get a light source at this wavelength in the form of a red laser pointer on Canal St. for about $1.00. Cy5 is soluble in water so you dissolve a 1 micro-molar (µM) concentration in a cylindrical sample bottle 1 inch in diameter. For a detector you use your iPhone camera and an App that records the intensity that the camera sees (App: LuxMeter). You do two experiments. First you measure the transmitted light intensity seen by the camera without Cy5 in the water, and then with Cy5 in the water. You find that the ratio of the recorded intensity with Cy5 to that without Cy5 is 0.5. What is the absorption cross section of Cy5 at the laser wavelength?

Exercise 3.11 – Find a way to get Beer's Law by counting up absorption sites in solution. Hint: Each absorption event eliminates power $\sigma \cdot I(x)$ from the beam. See Fig. 3.16.

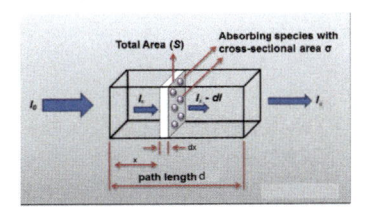

Figure 3.16 - Radiation from the laser is absorbed by the solution according to Beer's Law.

Experiment #4 – Spectrophotometry: Beer's Law and atomic absorption Cross-Section.

Your Pocket Optics kit includes experimental components for verifying Beer's Law. In addition to verifying Beer's Law by measuring the concentration dependence of the transmitted intensity through a $CuSO_4$ solution, you are asked to use your data to determine the absorption cross section of a Cu^{2+} ion at your laser wavelength. A sample squeeze bottle having 0.5M Cu^{2+} is included along with 4 separate empty cuvettes. This will allow you to prepare your own concentrations by dilution. By measuring the transmission of light through each sample you prepare using your Photometry Board, you should be able to verify the exponential attenuation derived in Eq. (3.57) and determine the absorption cross section of Cu^{2+}. With care you should be able to get 4 data points, with one being that for pure water. Shown below is my setup.

Scan for a video demo of this experiment

Experiment#4 Setup

Chapter Four – Polarization of Light

4.1 – Edwin Land and the Motivation for Polarized Spectacles

One of the great innovators of the 20th century was the founder of Polaroid Corporation, Edwin Land. Land developed an interest in polarized light while at a summer camp he attended as a high school student and carried this interest with him as he entered his undergraduate career at Harvard. After his freshman year in 1926 Land dropped out of Harvard to work on developing polarizers. He moved to New York City, studied in the New York Public Library, and snuck into laboratories at Columbia University to perform his research. After creating the sheet polarizer he returned to Harvard in 1929 and was given his own lab. He dropped out of Harvard a second time in 1932 to form Land-Wheelwright Labs, which later became the Polaroid Corporation.

Edwin Land was motivated to create a sheet polarizer in order to prevent glare from blinding automobile drivers and causing accidents, see Fig. 4.1. The value of understanding the generation of polarized light cannot be overstated. In Section 4.2 we will discuss the generation of polarized light by natural means, and then return to Land's engineered materials in Sec. 4.3.

Figure 4.1 – The glare off a roadway due to the reflection of sunlight can be greatly reduced by using polarizing spectacles. Since reflected sunlight from the roadway (glare) is polarized tangent to the roadway, in order to eliminate this hazard, spectacles are created which absorb this glare by orienting anisotropic molecules and sub-micron crystallites so that their absorbing axes are tangent to the roadway.

4.2 – Generation of Polarized light

In Section 1.4 we showed that a plane sheet of oscillating current generates plane-polarized radiation. In point of fact all one needs to produce polarized radiation is an oscillating charge. An individual oscillating charge does not produce plane waves. The radiation it produces is known as a spherical wave. Electromagnetic radiation can also be produced from a neutral oscillating dipole moment that is driven by an external wave. Recall that this sort of thing was discussed in Chap. 1 when describing optical tweezers. The dipole moment $\underline{\mu}(t) = \alpha \underline{E}(t)$, where α is the polarizability. What we didn't discuss at the time is the radiation pattern that emanates from this oscillating dipole, known as Rayleigh scattering. Fig. 4.2 shows the radiation fields from a vertically oriented oscillating dipole moment. The major things to notice are that the radiated field amplitude is null along the dipole axis and maximum perpendicular to this axis; the radiated intensity $I(\theta) \propto \sin^2 \theta$. These simple facts allow us to understand why radiation as seen overhead from our atmosphere is polarized.

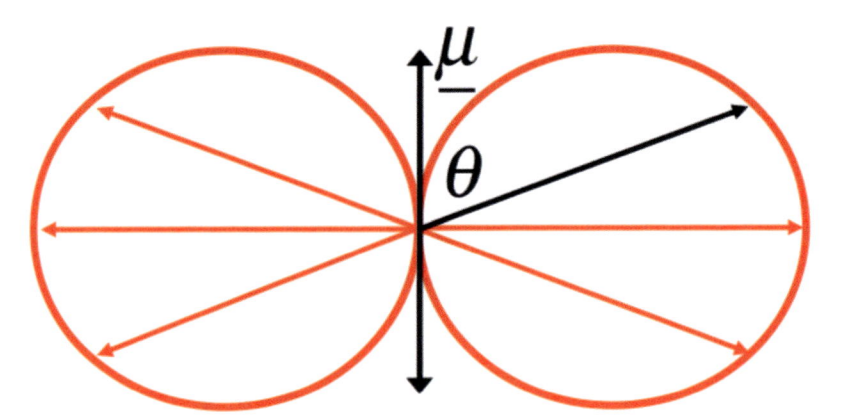

Figure 4.2 – Time averaged radiation intensity pattern from a vertically oscillating dipole moment, as represented by the length of the red arrows. The pattern of emitted intensity is in proportion to $\sin^2 \theta$ and is axially symmetric. It shows no radiation along the dipole direction, and maximum radiation perpendicular to the dipole direction.

Although light from the sun is un-polarized, when sunlight enters the atmosphere it induces oscillating dipoles in atmospheric molecules. From the illustration in Fig 4.3 one sees that the unpolarized light from sun generates polarized light for an observer looking overhead. The direction of this polarization is both perpendicular to the direction of the sunlight, and perpendicular to the line of sight. This polarization occurs because the field from the sun induces oscillating dipole moments in molecules. However a dipole cannot radiate along the axis of its dipole moment, so the induced polarization perpendicular to the line of sight is the optimal polarization direction for radiating downward.

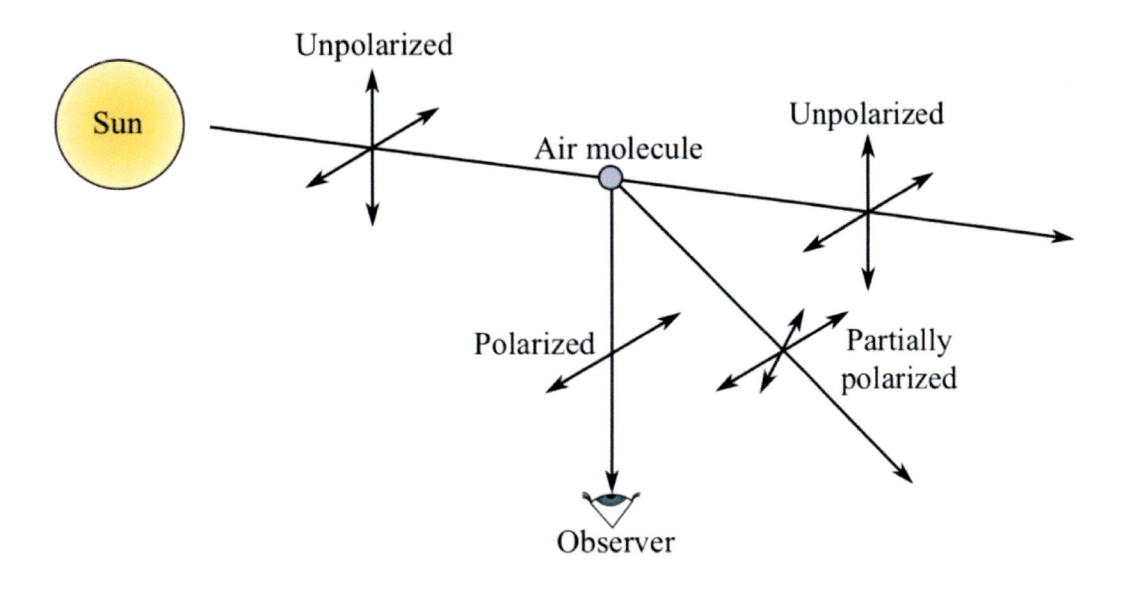

Figure 4.3 – The polarization of sunlight by air molecules.

Another means for polarizing light is Brewster reflection. Brewster reflection is associated with the loss of reflected energy for light polarized parallel to the plane of reflection, when the angle of incidence is equal to the Brewster angle. Consequently, unpolarized light incident at the Brewster angle generates a reflected beam that is polarized perpendicular to the plane of reflection, as illustrated in Fig. 4.4(a). Fig. 4.4(b), shows the angle of incidence dependence of the ratio R of reflected to incident intensity for polarized light, with one curve marked $R_\parallel$ for parallel polarization, and the other marked $R_\perp$ for perpendicular polarization. Note the total loss of the parallel polarized reflection at $56°18'$, for an air/glass interface for which $n_1 = 1.0$ and $n_2 = 1.5$. One hypothesis for this phenomenon is that the surface dipoles induced by the refracted light point in the direction of reflection at the Brewster angle, thereby eliminating reflection; the radiated field amplitude is null along the dipole axis. Next we will test this hypothesis.

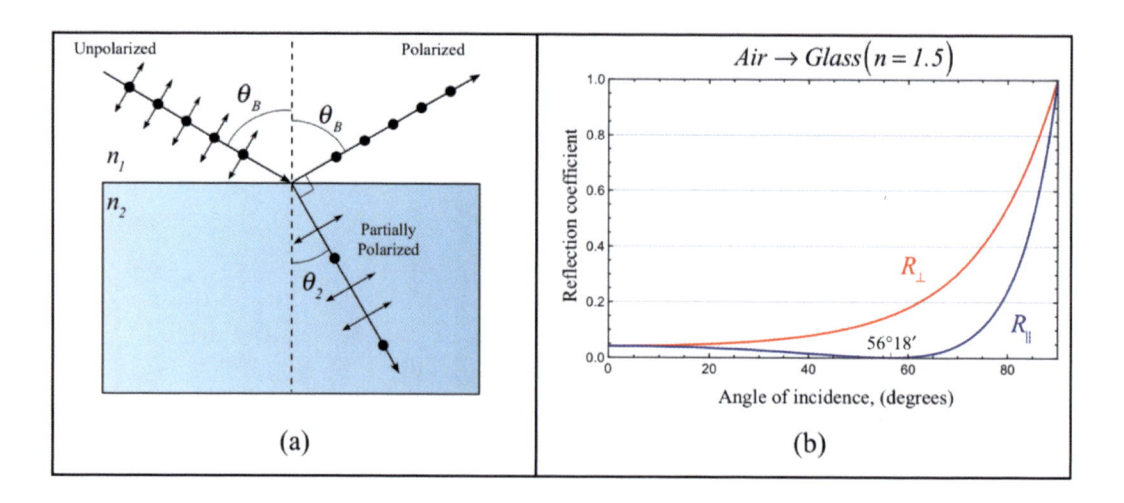

Figure 4.4 – The reflection of polarized light depends on angle of incidence. The parallel reflection coefficient $R_\parallel$ corresponds to the field polarized in the plane of the figure.

In Fig. 4.5 we suppose that only parallel polarized light is incident on a surface. First we recognize that induced dipoles at the surface are expected to be in the direction of the refracted light field, or perpendicular to the propagation direction of the refracted light. Therefore the angle from the normal to the induced dipole direction at the surface will be $\theta_d = \pi/2 - \theta_2$. If the incident angle were at the Brewster angle θ_B, based on our hypothesis the dipoles should align in the direction for which we would normally expect reflection to occur, and reflection would cease. At that point $\theta_1 = \theta_d = \theta_B$, and Snell's law would read

$$n_1 \sin(\theta_B) = n_2 \sin(\theta_2) = n_2 \sin(\pi/2 - \theta_B) = n_2 \cos(\theta_B),$$

or

$$\tan(\theta_B) = n_2/n_1. \qquad (4.1)$$

For the situation depicted in Fig. 4.4(b), $n_2/n_1 = 1.5$, for which Eq. (4.1) gives a value of $\theta_B = 56°18'$, in agreement with experiment. It should be pointed out that our approach to understanding the Brewster phenomenon is heuristic, however the full application of Maxwell's equations also yields Eq. (4.1), as we will see in the next chapter.

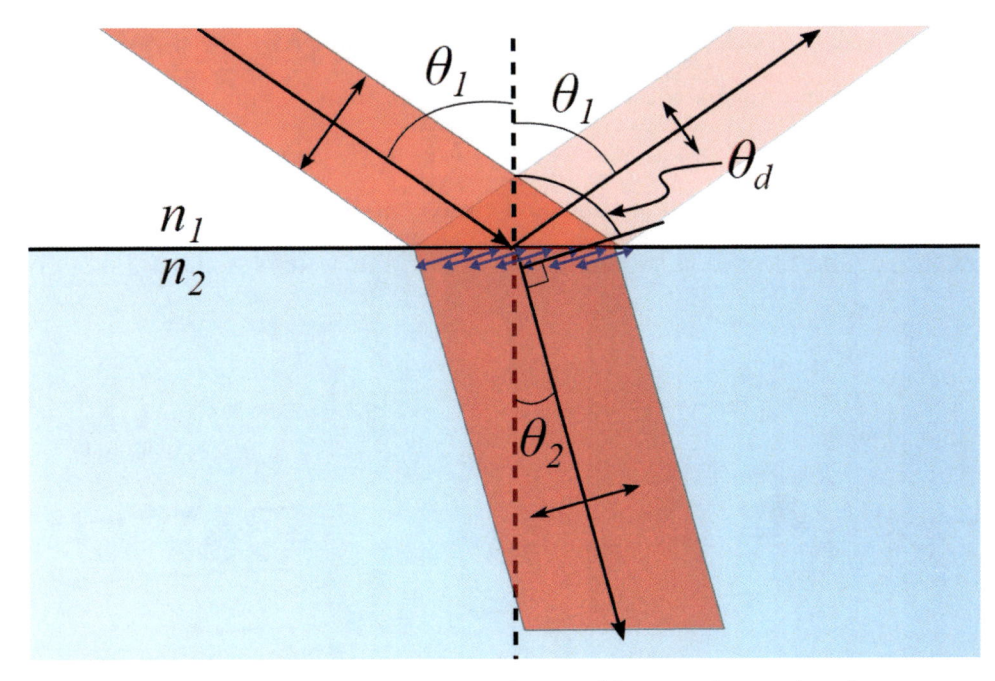

Figure 4.5 – The Brewster angle can be understood by postulating that the orientation of dipoles at the surface by parallel polarized light are in the direction of the field of the refracted wave.

Fig. 4.4 demonstrates that reflection from surfaces is principally polarized perpendicular to the plane of reflection, or tangent to the surface. So too is glare. E.H Land's quest was to engineer a material to eliminate this glare by absorbing the tangentially polarized scattering from the roadway surface, before reaching the eye.

4.3 – Linear Dichroism and E.H. Land's challenge

Another way to polarize unpolarized electromagnetic waves is to absorb energy from a field component in one direction while transmitting energy from the field component at right angles. A material with this property is known as dichroic. In the microwave region a dichroic polarizer is a grid of conducting wires in a frame. Figure 4.6 shows such a wire grid oriented along the x-direction for light propagating along the z-direction. The field component in the x-direction drives electrons in each of the wires. These electrons collide with lattice ions and give up energy to heat. The overall effect is that the component of the field polarized along the wires (x-direction) is not transmitted. On the other hand the component along the perpendicular y-axis interacts weakly with the wires and is largely transmitted. The wire grid is a dichroic polarizer that can only work if the wires are spaced considerable less than a wavelength apart. In the microwave region this corresponds to a wire spacing of less than 1 mm. Creating such a mechanism for visible light was the quest of Edwin H. Land. A major problem is that visible wavelengths are ten thousand times smaller than wavelengths in the microwave region. So we are talking about the grid of wires less than a tenth of a micron apart, 100 nm.

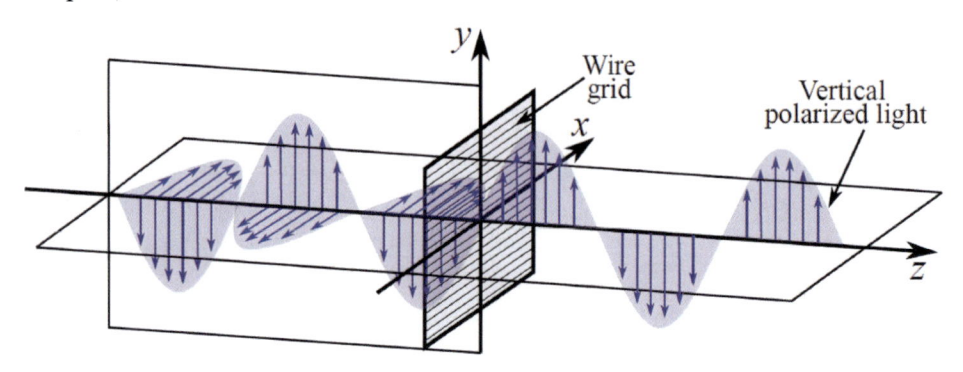

Figure 4.6 – The electric field of a wave having both vertical and horizontal polarization components is incident on a conducting grid of horizontal wires. Energy lost by the horizontally polarized component due principally to ohmic losses does not allow it to be transmitted.

The materials challenge that Edwin Land took on as he searched for such a material at the NY public library in the late 1920's was daunting (Land, 1951). Ultimately he came upon a crystal known as Herapathite. It absorbed light polarized along one axis, and was almost transparent along a perpendicular axis. William Bird Herapath discovered Herapathite in 1852 (Herapath, 1852). The story of its chance discovery involves a student of Herapath who noticed that while adding iodine to the urine of a quinine-fed dog, unusual green crystals were formed. Under a microscope Herapath observed that these crystals polarized light. Land's discovery of this early work led him to anticipating the growth of large crystals of Herapathite. Although that proved to be impractical, he decided to disperse sub-micron sized needle shaped crystals on a plastic sheet, and orient them both mechanically, and with electric and magnetic fields. Land dubbed this first large area polarizer based on Herapathite the J-sheet.

The Herapathite unit cell gets its one-dimensional dichroic nature by electronic conduction along a series of iodine triads, I_3^-, effectively producing conducting nano-wires on the order of one hundredth of a micron apart (Kahr, *et al*, 2009). The problem with using these needle shaped crystals goes beyond the difficulty in orienting them. Small crystals with high index contrast with their surroundings and scatter light. This leads to a sheet of material through which images are not particularly clear. To overcome this problem, Land turned to long chain polymers. In particular he found in the case of polyvinyl alcohol that he could align the polymer chains by stretching the sheet along one axis while heating it. By bathing this normally insulating sheet in an iodine solution, the chains that he referred to as "Polymeric Iodine" became conductive, just as in the case of Herapathite. With no discrete micro-crystals, however, the scattering problem was eliminated, and clear images were seen through this large-area polarizing sheet. This product was known as the H-sheet.

The H-sheet is illustrated in Fig. 4.7 as a sheet of molecular nano-wires all oriented horizontally. Light polarized horizontally will be absorbed, while light polarized along the transmission axis $\hat{d}$ will only be slightly attenuated.

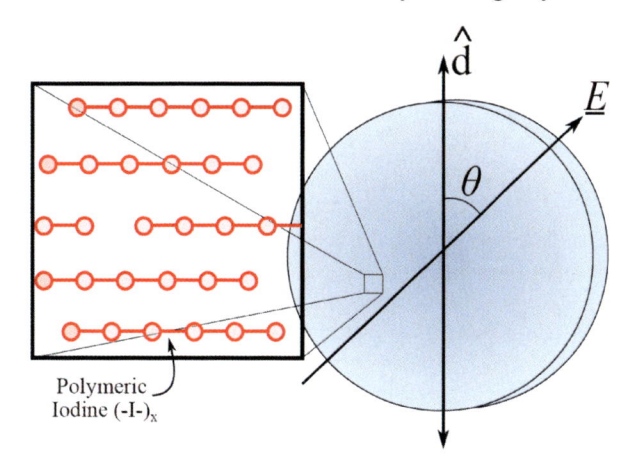

Figure 4.7 – A polarizer built from "Polymeric Iodine". In this case, the transmission axis is vertical, so the electrons within its "zillion" cages can only move horizontally.

The transmission of light in Fig.4.7 only depends on the angle between the field of the incident light and the transmission axis, θ. Ideally the transmission is 100% when $\underline{E}$ and $\hat{d}$ are parallel, whereas at $\theta = 90°$ absorption is so strong that transmitted light ceases. Between these two extremes the transmitted intensity $I(\theta)$ can be obtained by considering the component of the incident field along the transmission axis. If the incident light field is $\underline{E}(t) = \underline{E}_0 e^{-i\omega t}$, then its component along the transmission axis is $\underline{E}(t) \cdot \hat{d}$, so that the ratio of the time averaged intensity represented by this field component to the time averaged intensity of the incident light is $\left|\underline{E}(t)\cdot\hat{a}\right|^2 / E_0^2 = \cos^2\theta.$ In terms of intensity

$$I(\theta) = I(0)\cos^2(\theta). \tag{4.2}$$

This equation is known as the Law of Malus.

The incident light directed into the page in Fig. 4.7 is linearly polarized at angle θ from the transmission axis. Light directly from the sun is randomly polarized, or unpolarized. At any time the direction of the electric field can point in any direction in the plane of the polarizer. Because this light has an equal probability to be at any angle θ, one must average over θ to determine the fraction of unpolarized light that gets through the dichroic polarizer;

$$\frac{1}{2\pi}\int_0^{2\pi}\cos^2(\theta)\,d\theta = \frac{1}{2},$$

so half of the incident light is absorbed. The light transmitted through the polarizer will be linearly polarized along the transmission axis.

4.4 – Birefringence and the Generation of Circularly Polarized Light

When discussing the real part of the refractive index in Chapter 3 we allowed the caged electron to absorb light polarized in a single direction. A more realistic model would allow motion along all 3 axes. For nanoscopic objects, dealing with motion in more than one direction can have interesting consequences with respect to the change of the refractive index with polarization direction. To understand this we will explore a modified version of the Lorentz atom as shown in Fig. 4.8, and imagine having a medium built of many of these atoms, all with identical orientations. Treating the direction of light propagation along the z-axis, we see the springs along the x and y axes have different strengths as indicated by their thickness. Thus we assign the springs different constants and therefore there are different resonance frequencies depending on the direction of electron oscillation. This results in a refractive index that depends on the direction of the electric field polarization as it interacts with the electron cloud. In particular, the frequency of resonance for y polarized light is higher than that for x polarized light, due to the stiffer springs along the y direction. That means that at a light frequency ω_b the refractive index will be higher for x versus y polarized light as illustrated. This Lorentz atom system will therefore be birefringent at ω_b (i.e. subscript b for birefringent). A major consequence of birefringence is the generation of circularly polarized light.

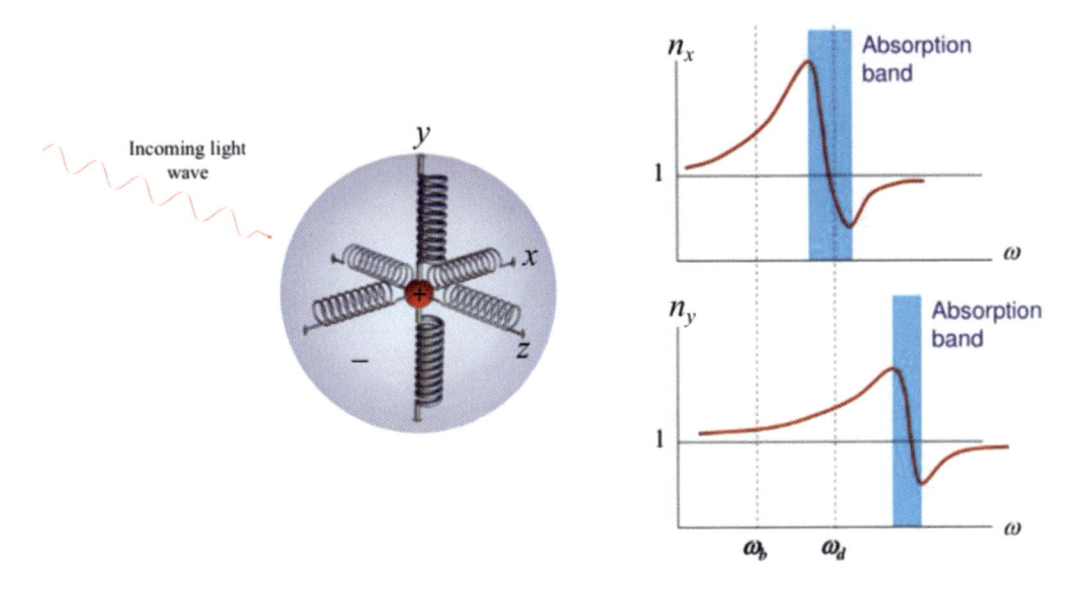

Figure 4.8 – A Lorentz oscillator for which the spring constant is different depending on the direction of oscillation. ω_b stands for the frequency at which a material made of these Lorentz oscillators demonstrates birefringence, whereas the same material will demonstrate dichroism at ω_d.

Circular polarized light is an electromagnetic wave in which the electric field coming to an observer rotates with time. If it rotates clockwise it is called "right circular polarized" or RCP, counter to this it is called "left circular polarized", LCP. Circular polarized light is usually generated with a quarter wave plate ($\lambda/4$). A $\lambda/4$ plate is typically cut from a quartz crystal in such a way that light polarized in one direction has a lower refractive index (n_f) than that in a perpendicular direction (n_s). These two directions are known as fast f and slow s directions since the light develops a larger velocity within the quartz when polarized along the f direction in comparison with the s direction. To generate circularly polarized light, linearly polarized light incident on the plate is directed at 45° to the fast axis so that the electric field components along the s and f directions are the same (Fig. 4.9). The plate is cut to a thickness such that the s transmitted wave component is 90° out of phase with the f polarized wave

115

component. This phase difference arises from the difference in refractive indices n_f and n_s. To understand this let's think about how the angular spatial frequency of light k changes when it travels thru a medium with a different refractive index. Recall the wave vector changes from its free space value k_0 by simply multiplying by the refractive index; $k = nk_0$. This can be shown as follows,

$$k = \frac{2\pi}{\lambda} = \frac{2\pi}{(c/n)/f} = \frac{2\pi n}{\lambda_0} = nk_0. \qquad (4.3)$$

When a wave travels across a plate of thickness d the amount of phase covered from the front side of the plate to the backside is $nk_0 d$. Since the f and s waves start out with the same phase on the front side they finish by different phases

$$\Delta\phi = (k_s - k_f)d = \frac{2\pi(n_s - n_f)d}{\lambda_0} \qquad (4.4)$$

on the backside.

For a quarter-wave plate we would like $\Delta\phi = \pi/2$, in order to generate circularly polarized light from the incident linearly polarized light. From Eq. (4.4) this requires a thinnest plate thickness of

$$d = \frac{\lambda_0}{4(n_s - n_f)} \qquad (4.5)$$

The net effect of this plate thickness is to cause the reconstructed wave polarization to propagate in a helical pattern as shown in Fig. 4.9. An observer looking in from the right side of this illustration ideally would see the electric field rotating clockwise; RCP.

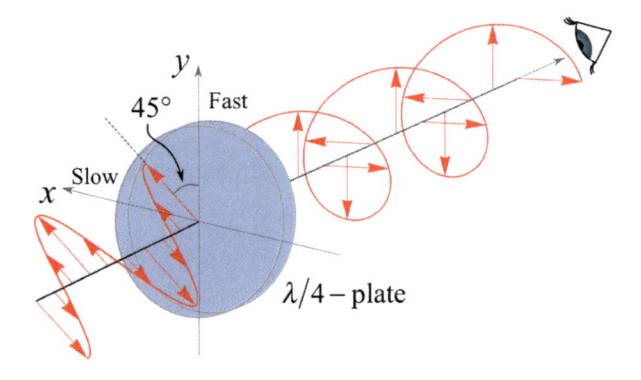

Figure 4.9 – Conversion from linear to circularly polarized light using a quarter-wave plate.

Quantum Optics point of view with respect to circular polarization

According to Quantum Optics pure Circular Polarized light is light with photons all in the same state of angular momentum. Each photon carries an angular momentum component parallel or anti-parallel to the propagation direction with magnitude $\hbar = h/2\pi$, where h is Planck's constant. For LCP light the angular momentum of the photons is in the direction of propagation $\hat{k}$; $\underline{L} = +\hbar\hat{k}$, see top of Fig. 4.10. Photons in the opposite state are right circular polarized, RCP and have angular momentum $-\hbar\hat{k}$ opposite to the direction of propagation (bottom of Fig. 4.10).

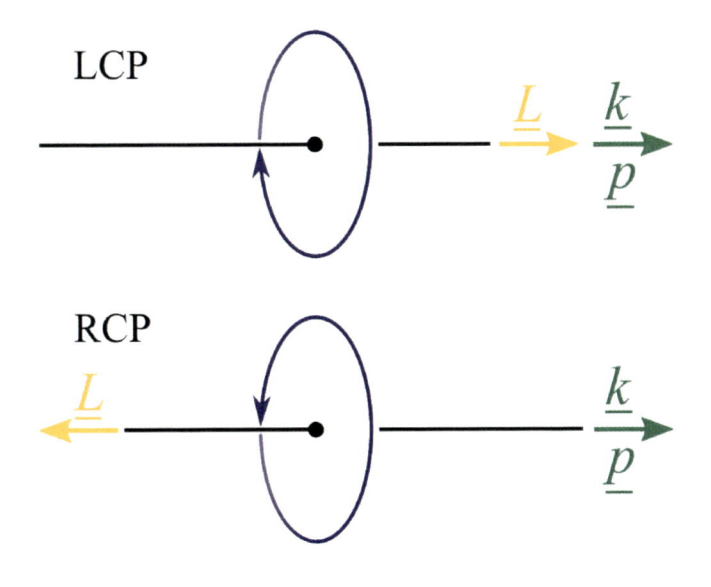

Figure 4.10 – Angular momentum of a photon.

4.5 – Polarization and Jones Calculus

We would like to develop a systematic way to describe polarization states of light and optical devices that manipulate polarization (e.g. polarizer, $\lambda/4$ plate, $\lambda/2$ plate). It turns out that polarization states can be represented conveniently using vectors, and the optical devices used to change polarization states can be described with matrices. Let's begin by developing polarization vectors. The polarization vector of a plane light wave is always perpendicular to the direction of propagation of the wave. By choosing the z-axis (out of the page in Fig. 4.11) as the direction of propagation we restrict the polarization vector to the *xy*-plane.

If the wave is simply polarized horizontally or vertically then at an instant in time it looks like the upper or lower vector illustrations shown on the left in Fig. 4.11. On the right the polarization is represented by a column vector, with the upper component being the horizontal (*x*) and the lower being the vertical (*y*). The complex time dependence $e^{-i\omega t}$ is written for convenience with the understanding that the real part is what counts; a horizontally polarized field could also have been written as $E_0 \hat{x} \cos(\omega t)$. The polarization vector of light need not be confined to a line, in some cases the polarization vector traces out a circle. This is called circularly polarized light.

$$E_h = E_0 e^{-i\omega t} \begin{bmatrix} 1 \\ 0 \end{bmatrix}$$

$$E_v = E_0 e^{-i\omega t} \begin{bmatrix} 0 \\ 1 \end{bmatrix}$$

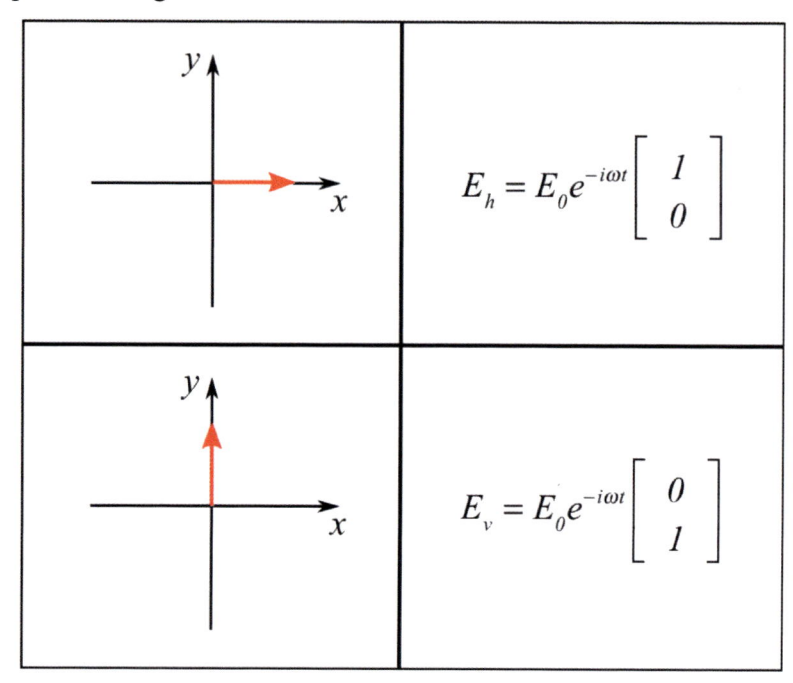

Figure 4.11 – Horizontal (upper) and vertical (lower) linear polarized light.

Circularly polarized light results from a phase shift between the *x* and *y* components of the field. Right circular polarized light (RCP) is represented by a field that rotates in the x-y plane in the clockwise direction, as shown in the upper left hand diagram in Fig. 4.12. The mathematical representation on the upper right now has an added phase in the *y*-component. Although the *x*-component is proportional to the $\cos(\omega t)$, the y-component is proportional to $\cos(\omega t + \pi/2)$. That means that at $t = 0$ the field is along the x-axis, whereas at $t = \pi/(2\omega)$ the field is along the negative *y* direction; as time increases the field circles clockwise. The wave approaching is right circularly polarized (RCP or R). The opposite is true in the lower figure and equation in Fig. 4.12 here the wave approaching is left circularly polarized ("LCP", or L).

$$E_R = \frac{E_0}{\sqrt{2}} e^{-i\omega t} \begin{bmatrix} 1 \\ e^{-i\frac{\pi}{2}} \end{bmatrix}$$

$$E_L = \frac{E_0}{\sqrt{2}} e^{-i\omega t} \begin{bmatrix} 1 \\ e^{i\frac{\pi}{2}} \end{bmatrix}$$

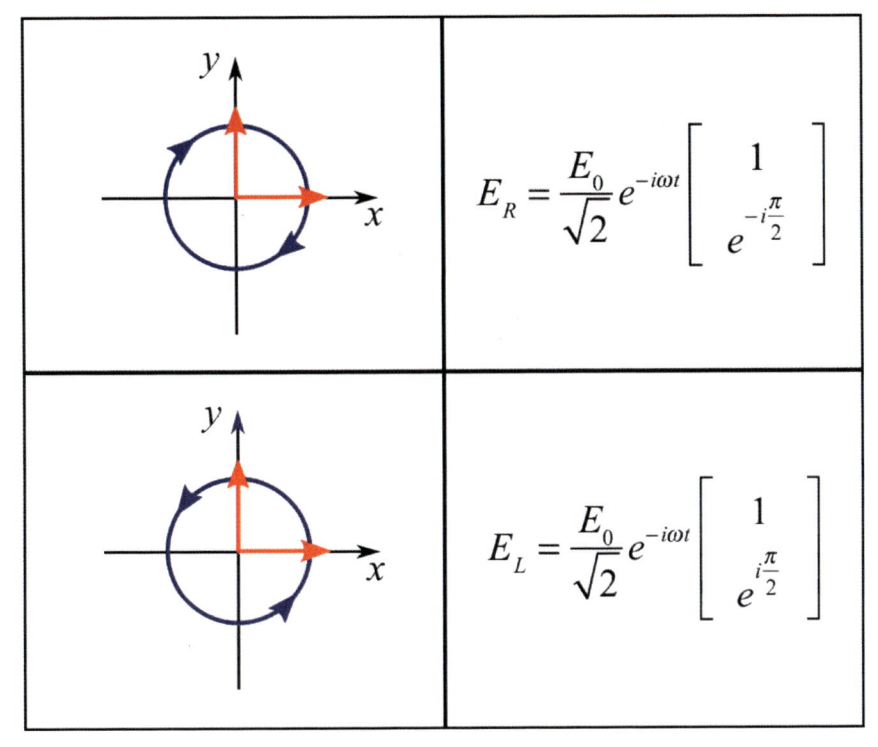

Figure 4.12 – Representations of RCP (upper) and LCP (lower) light.

One thing that may be confusing is that a factor of $1/\sqrt{2}$ appears in the expressions for the circularly polarized light in Fig. 4.13 but not for the linearly polarized expressions in Fig. 4.12. This is because each is normalized to the same intensity, which for complex vectors is expressed as $E^\dagger E = E_0^2$. The dagger stands for complex conjugate transpose. To demonstrate the dagger operation on the RCP field in Fig. 4.12 we turn the column vector into a row and take the complex conjugate;

$$E_R^\dagger = \frac{E_0}{\sqrt{2}} e^{i\omega t} \begin{bmatrix} 1 & e^{i\frac{\pi}{2}} \end{bmatrix}, \qquad (4.6)$$

so

$$E_R^\dagger E_R = \frac{E_0^2}{2} \begin{bmatrix} 1 & e^{i\frac{\pi}{2}} \end{bmatrix} \begin{bmatrix} 1 \\ e^{-i\frac{\pi}{2}} \end{bmatrix} = \frac{E_0^2}{2}(1+1) = E_0^2 \qquad (4.7)$$

All of our base fields (E_R, E_L, E_h, E_v) will be normalized to this value.

119

Now we are ready to construct the mathematics of different polarization operators (e.g. polarizer, $\lambda/4$ plate, $\lambda/2$ plate). The simplest of all of these is a linear polarizer along the x or y-axis. For the x oriented polarizer the device should allow all of the intensity of a horizontal polarized beam to pass through, but only half of a circularly polarized beam, because the field in a circularly polarized beam can have all directions in the x-y plane with equal probability. The correct solution to this problem is the following operator

$$\underline{\underline{A}}_{Lh} = \begin{bmatrix} 1 & 0 \\ 0 & 0 \end{bmatrix}. \tag{4.8}$$

When operating with this on a horizontally polarized beam we get

$$\underline{\underline{A}}_{Lh} E_h = E_0 e^{-i\omega t} \begin{bmatrix} 1 & 0 \\ 0 & 0 \end{bmatrix} \begin{bmatrix} 1 \\ 0 \end{bmatrix} = E_0 e^{-i\omega t} \begin{bmatrix} 1 \\ 0 \end{bmatrix}, \tag{4.9}$$

which is what we started with. However when operating with this same horizontal polarizer on a vertical polarized beam,

$$\underline{\underline{A}}_{Lh} E_v = \begin{bmatrix} 1 & 0 \\ 0 & 0 \end{bmatrix} E_0 e^{-i\omega t} \begin{bmatrix} 0 \\ 1 \end{bmatrix} = E_0 e^{-i\omega t} \begin{bmatrix} 0 \\ 0 \end{bmatrix}, \tag{4.10}$$

we loose both field components; no light comes through. Let's try sending a normalized RCP beam through this polarizer

$$\underline{\underline{A}}_{Lh} E_R = \begin{bmatrix} 1 & 0 \\ 0 & 0 \end{bmatrix} \frac{E_0}{\sqrt{2}} e^{-i\omega t} \begin{bmatrix} 1 \\ e^{-i\frac{\pi}{2}} \end{bmatrix} = \frac{E_0}{\sqrt{2}} e^{-i\omega t} \begin{bmatrix} 1 \\ 0 \end{bmatrix}. \tag{4.11}$$

This time the circularly polarized beam is turned into a linearly polarized beam, but the amplitude is reduced by the square root of 2, and consequently the intensity is halved, just as we expected from the beginning. There are clearly other cases that can be tried. You should ask about the operator for a vertical polarizer (see Exercise 4.3). Next we will construct a circular polarizer.

A $\lambda/4$ plate turns linear to circularly polarized light when linearly polarized light is oriented to have equal components in the horizontal and vertical directions (Fig. 4.13).

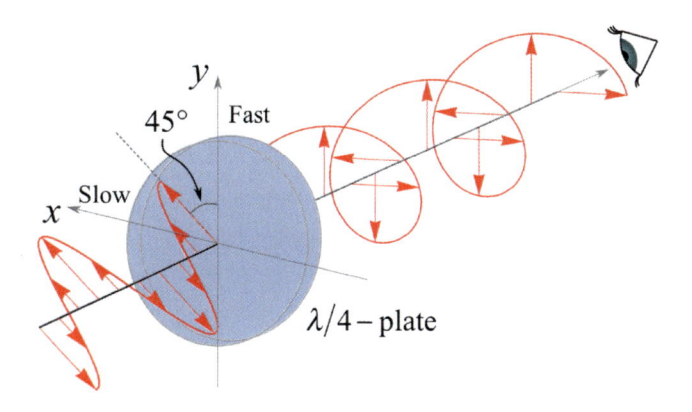

Figure 4.13 – Right circularly polarized light (RCP) produced from linear polarized light by using a quarter-wave plate.

In Fig. 4.13, light propagates from left to right. To describe the plane-polarized light hitting the circular polarizer at its left face we require equal horizontal and vertical components. The light field is therefore

$$E_{45} = \frac{E_0}{\sqrt{2}} e^{-i\omega t} \begin{bmatrix} 1 \\ 1 \end{bmatrix}. \tag{4.12}$$

Note the square root of 2, which makes $E^\dagger E = E_0^2$. A quarter wave plate will phase shift one component more than the other by 90° in order to generate RCP light. To get from Eq. (4.12) for E_{45} to E_R will require a $\lambda/4$ plate,

$$\underline{\underline{A}}_{\lambda/4} = \begin{bmatrix} 1 & 0 \\ 0 & e^{-i\frac{\pi}{2}} \end{bmatrix}. \tag{4.13}$$

If you operate with this $\lambda/4$ plate on E_{45},

$$\underline{\underline{A}}_{\lambda/4} E_{45} = \frac{E_0 e^{-i\omega t}}{\sqrt{2}} \begin{bmatrix} 1 & 0 \\ 0 & e^{-i\frac{\pi}{2}} \end{bmatrix} \begin{bmatrix} 1 \\ 1 \end{bmatrix} = \frac{E_0 e^{-i\omega t}}{\sqrt{2}} \begin{bmatrix} 1 \\ e^{-i\frac{\pi}{2}} \end{bmatrix}, \tag{4.14}$$

you get RCP light. To get a half wave plate we combine two $\lambda/4$ plates,

$$\underline{\underline{A}}_{\lambda/2} = \underline{\underline{A}}_{\lambda/4} \underline{\underline{A}}_{\lambda/4} = \begin{bmatrix} 1 & 0 \\ 0 & e^{-i\frac{\pi}{2}} \end{bmatrix} \begin{bmatrix} 1 & 0 \\ 0 & e^{-i\frac{\pi}{2}} \end{bmatrix} = \begin{bmatrix} 1 & 0 \\ 0 & e^{-i\pi} \end{bmatrix} = \begin{bmatrix} 1 & 0 \\ 0 & -1 \end{bmatrix}. \tag{4.15}$$

Operating on E_R we expect to get E_L;

$$\underline{\underline{A}}_{\lambda/2} E_R = \begin{bmatrix} 1 & 0 \\ 0 & -1 \end{bmatrix} \frac{E_0}{\sqrt{2}} e^{-i\omega t} \begin{bmatrix} 1 \\ e^{-i\frac{\pi}{2}} \end{bmatrix} = \frac{E_0}{\sqrt{2}} e^{-i\omega t} \begin{bmatrix} 1 \\ -e^{-i\frac{\pi}{2}} \end{bmatrix} = \frac{E_0}{\sqrt{2}} e^{-i\omega t} \begin{bmatrix} 1 \\ e^{-i\frac{3\pi}{2}} \end{bmatrix} = \frac{E_0}{\sqrt{2}} e^{-i\omega t} \begin{bmatrix} 1 \\ e^{i\frac{\pi}{2}} \end{bmatrix}. \tag{4.16}$$

4.6 – Chapter Four Exercises

Exercise 4.1 – If a LCP photon encounters a $\lambda/2$ plate initially at rest in the coordinate system of an observer in space, what will happen to the plate after the encounter ?

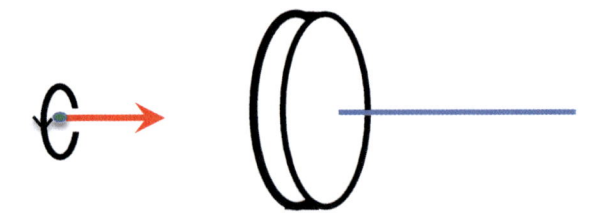

Hint – Circularly polarized light reverses polarization upon transmitting through a half-wave plate.

Exercise 4.2 – Calcite is a birefringent crystal with $n_f = 1.4864$ and $n_s = 1.6584$.

In order to circularly polarize linearly polarized light of wavelength $\lambda = 590\ nm$ an experimenter cuts a quarter wave-plate out of calcite. What thickness should he cut the crystal for the thinnest quarter wave plate?

Exercise 4.3 – Construct an operator (Jones matrix) for a vertical polarizer.

Exercise 4.4 – Construct a normalized linear polarized field at an arbitrary angle θ to the horizontal axis.

Exercise 4.5 – A horizontal polarizer is placed in series with a vertical polarizer. What is the total transmittance of the system for incident unpolarized light?

Exercise 4.6 – Now place an additional polarizer between the two polarizers (Ex. 4.5) at a 45° angle relative to the horizontal. What is the total transmittance? What happens if you place the 45° polarizer before or after the two perpendicular polarizers? Does changing the order of the polarizers change the transmittance? Why or why not?

Exercise 4.7 – Linear polarized laser radiation of intensity I_0 with its electric field polarized in the y direction is incident on a dichroic plate with its transmission axis ($\hat{d}$) at 45° to the x and y axes (as shown). The $\lambda/4$ plate following this dichroic plate has its fast axis along y and its slow axis along x (as shown).

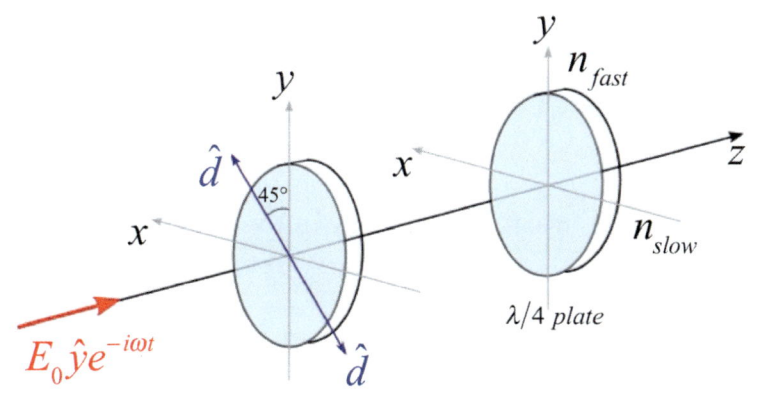

(a) Using the Law of Malus calculate the intensity in terms of I_0 for the light passing between the dichroic plate and quarter wave plate.

(b) Using Jones calculus identify the polarization state and calculate the intensity of the light (in terms of I_0) passing through the quarter wave plate.

Experiment #5 – Laser Polarization and Law of Malus

In your Pocket Optics kit you will find a white cylindrical sleeve that fits over your laser and easily rotates. The sleeve has affixed to one side a polarizer. As you rotate the sleeve it modulates the intensity of the laser. By measuring the laser intensity as a function of the angle of rotation you should be able to verify the Law of Malus if your laser is linearly polarized. As before power readings can be taken from the numerical display. My set-up is shown below.

Scan for a video demo of this experiment

Experiment#5 Setup

Pedagogy – Section 4.3, and Eq. (4.2)

Chapter Five – Theory of Reflection and Refraction at an Interface

5.1 – Reflection and Transmission Coefficients

Now we will deal with the effects that refractive index has with respect to reflection from surfaces. There are three important questions we would like to answer about reflection: (1) how much light is reflected from a dielectric surface and how does it depend on refractive index? (2) how do wave phases change in reflection and transmission? (3) How does reflection from a dielectric differ from reflection from a metal?

We start with a simple situation shown in Fig. 5.1. Light approaches a dielectric surface (e.g. glass or plastic) from the left. Upon interaction the light is both reflected and transmitted. Of course a single photon can only do one or the other, however we will deal with wave theory in trying to understand reflection and transmission. In wave theory we write out three waves: an incident wave (left to right), a reflected wave (right to left), and a transmitted wave (left to right).

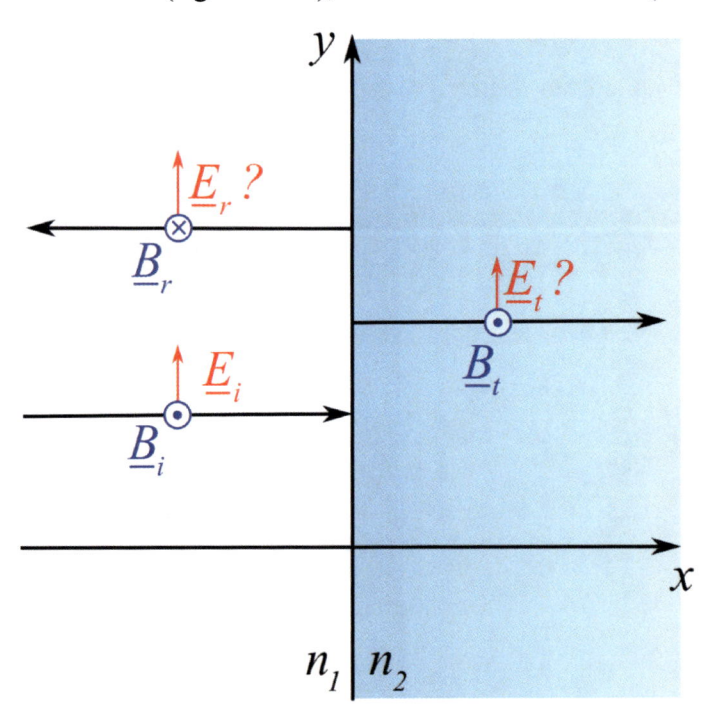

Figure 5.1 – Reflection and transmission at a dielectric boundary. Note that we have picked the polarization directions for all waves to be simultaneously upward at the boundary. We only need to pick the incident polarization, Maxwell's equations will tell us whether the other polarizations are correct.

Our choice for polarization of the fields at the surface may seem unwise if you remember from your introductory physics that reflection of light arriving from air by a glass surface should lead to a reversal in the reflected field, however we should not need to know this in advance; consequently you will notice the question marks in Fig.5.1. If we fix the incident polarization at the surface to be up at the instant shown, then Maxwell's theory should tell us if the reflected and field is in the correct direction. For that reason question marks have been placed behind our choices for the polarizations of the reflected and transmitted fields. Since Maxwell's equations are linear we expect the amplitude of the reflected and transmitted waves to be proportional to the amplitude of the incident wave, so that at time $t=0$ each of the three waves can be written as:

$$\underline{E}_i = E_{i0}\hat{y}e^{+in_1k_0x}, \tag{5.1}$$

$$\underline{E}_r = rE_{i0}\hat{y}e^{-in_1k_0x}, \text{ and} \tag{5.2}$$

$$\underline{E}_t = tE_{i0}\hat{y}e^{+in_2k_0x}. \tag{5.3}$$

Note the use of reflection and transmission coefficients, r and t, to describe the amplitudes of the reflected and transmitted waves, and the positive and negative exponents to account for the directions of the waves (i.e. negative for the reflected wave since it is moving in the negative x-direction). Also, note the change in refractive index for the transmitted wave in comparison to the incident and reflected waves. The goal now is to determine how r and t depend on the physical properties of the boundary (i.e. the refractive indices n_1 and n_2).

Our goal will require generating two simultaneous linear equations containing both r and t by relating the fields at the boundary of Fig. 5.1. The first boundary rule is simple: the total tangential electric field on the right hand side of the boundary should equal the total tangential electric field on the left hand side of the boundary. This boundary condition is arrived at from Faraday's law. To understand this, imagine taking the path integral over electric field using a path that hugs the boundary while looping around it in the xy-plane, see Fig. 5.2.

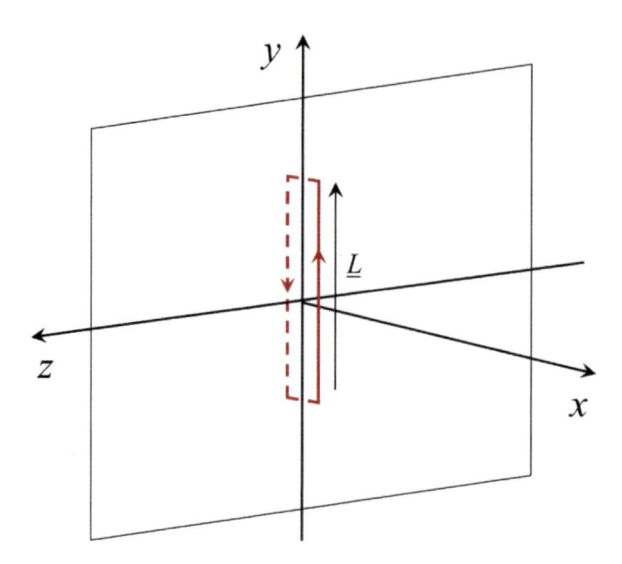

Figure 5.2 – Path integral for evaluating Faraday's Law.

The area of the loop is so small that it does not allow any magnetic flux to thread through it. Thus the path integral is equal to zero;

$$\oint_\ell \underline{E} \cdot d\underline{s} = 0. \tag{5.4}$$

This is the equivalent to Kirchhoff's circuit law. It is easy to implement in the current case since all of the electric fields are tangential to the boundary. Evaluating the integral gives

$$\underline{E}_t \cdot \underline{L} - \underline{E}_i \cdot \underline{L} - \underline{E}_r \cdot \underline{L} = 0 \tag{5.5}$$

From our original equations for the fields and because the boundary is at $x = 0$, the application of the boundary condition gives

$$E_{i0} + rE_{i0} = tE_{i0} \text{ or} \tag{5.6}$$

$$1 + r = t. \tag{5.7}$$

To determine how r and t are controlled by the physical properties of the boundary we will require another boundary condition.

The second boundary condition involves the magnetic field. Here for a non-magnetic material there is a similar boundary condition: the total tangential magnetic field on the right hand side of the boundary should equal the total tangential magnetic field on the left hand side of the boundary. Of course we will need the magnetic fields that belong to each of the waves. Our picks for the direction of the magnetic field for each of the waves are evident in Fig. 5.1, but may not be obvious. They are based on the so-called Poynting vector. Referred to as $\underline{S}$, the Poynting vector is in the direction in which the wave travels and has a magnitude equal to the instantaneous intensity carried by the wave. It is described in a non-magnetic medium by the following cross-product,

$$\underline{S} = \frac{\underline{E} \times \underline{B}}{\mu_0} \qquad (5.8)$$

For example, if at an instant the electric field of the incident wave is in the $+y$ direction, then the magnetic field $\underline{B}$ would have to be in the $+z$ direction in order for the wave to travel in the $+x$ direction. Consequently, by referring to Eq. (5.1) the incident $\underline{B}$ field is

$$\underline{B}_i = +\frac{E_{i0}}{v_1} \hat{z} e^{+in_1 k_0 x} . \qquad (5.9)$$

If on the other hand we take the reflected wave to travel in the $-x$ direction with an instantaneous electric field in the $+y$ direction, the $\underline{B}$ field will be in the $-z$ direction

$$\underline{B}_r = -r\frac{E_{i0}}{v_1} \hat{z} e^{-in_1 k_0 x} . \qquad (5.10)$$

The transmitted $\underline{B}$ field will have the same polarization direction as the incident $\underline{B}$ field although the wave velocity will now be v_2,

$$\underline{B}_t = t\frac{E_{i0}}{v_2} \hat{z} e^{+in_2 k_0 x} . \qquad (5.11)$$

After applying the boundary condition on the tangential components of the $\underline{B}$ field we obtain a second linear equation in r and t,

$$\frac{E_{i0}}{v_1} - r\frac{E_{i0}}{v_1} = t\frac{E_{i0}}{v_2} \quad \text{or} \qquad (5.12)$$

$$1 - r = \frac{v_1}{v_2} t . \qquad (5.13)$$

With the recognition that $v_1/v_2 = n_2/n_1$, Eq. (5.7) and Eq. (5.13) may now be solved simultaneously in order to obtain r and t,

$$r = \frac{n_1 - n_2}{n_1 + n_2} \quad \text{and} \qquad (5.14)$$

$$t = \frac{2n_1}{n_1 + n_2}. \quad (5.15)$$

These equations are field coefficients. Note that since glass has a higher refractive index than air, r will be negative; the reverse of our original choice. On the other hand t will be positive; as chosen.

Next we want the intensity coefficients. Considering that the Poynting vector, $\frac{\underline{E} \times \underline{B}}{\mu_0}$, represents the instantaneous intensity, each of the time averaged intensities can be evaluated from $(n\varepsilon_0 c/2) E_y^* \cdot E_y$, which means that the intensity of the reflected beam is just r^*r times the intensity of the incident beam. Thus the fraction of the incident power that is reflected is $R = r^*r$.

Our analysis should have conserved energy. To test this we will add the intensity in the reflected and transmitted beams,

$$\frac{n_1 \varepsilon_0 c}{2}(r^*r)(E_i^* E_i) + \frac{n_2 \varepsilon_0 c}{2}(t^*t)(E_i^* E_i) = \frac{\varepsilon_0 c}{2}\left(n_1(r^*r) + n_2(t^*t)\right) E_i^* E_i. \quad (5.16)$$

By substituting the results for r and t from Eq. (5.14) and Eq. (5.15) into Eq. (5.16) the right hand side can be shown to be the intensity of the incident beam (Exercise 5.3), thereby verifying that our analysis satisfies conservation of energy. As a consequence of energy conservation, the fraction of the incident power that is transmitted, T, is the difference between the incident power and the fraction of the power that is reflected;

$$T = 1 - R = (n_2/n_1) t^* t. \quad (5.17)$$

5.2 – Snell's Law and Brewster's Angle

In the previous section we found a way to determine reflection and transmission coefficients for light that is incident normal to a surface. We would like to develop a more general theory that allows light to be incident on a surface at an arbitrary angle. This scenario is illustrated in Fig. 5.3 for incident electric field polarization in the plane of reflection (often referred to as Transverse Magnetic or TM polarization).

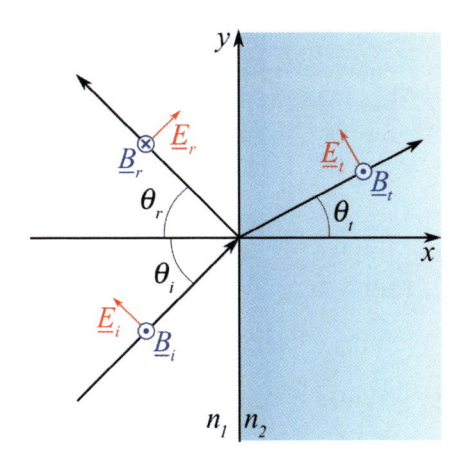

Figure 5.3 – Off normal incidence for the electric field parallel to the plane of reflection.

In Fig. 5.3 we are off normal incidence, but the sign conventions remain consistent with Fig. 5.1. Fig. 5.3 shows reflection and transmission for parallel incident polarization [i.e. the electric field is parallel to the plane of reflection (xy-plane)].

The boundary conditions are the same as before [i.e. the electric field parallel to the boundary ($x = 0$) on either side must be the same];

$$\underline{E}_i \cdot \hat{y} e^{i(\underline{k}_i \cdot y\hat{y} - \omega t)} + \underline{E}_r \cdot \hat{y} e^{i(\underline{k}_r \cdot y\hat{y} - \omega t)} = \underline{E}_t \cdot \hat{y} e^{i(\underline{k}_t \cdot y\hat{y} - \omega t)} . \qquad (5.18)$$

With the time phase divided out,

$$\underline{E}_i \cdot \hat{y} \cos(k_{iy} y) + \underline{E}_r \cdot \hat{y} \cos(k_{ry} y) = \underline{E}_t \cdot \hat{y} \cos(k_{ty} y) . \qquad (5.19)$$

For Eq. (5.19) to hold for all y

$$k_{iy} = k_{ry} = k_{ty} . \qquad (5.20)$$

The Eq. (5.20) gives rise to two familiar results. The first comes from $k_{iy} = k_{ry}$, which written in terms of angles in Fig. 5.3 is

$$k_i \sin\theta_i = k_r \sin\theta_r . \qquad (5.21)$$

Because the incident and reflected waves travel in the same medium $k_i = k_r$, so for Eq. (5.21) to hold

$$\theta_i = \theta_r ; \qquad (5.22)$$

the law of reflection. The second result comes from $k_{iy} = k_{ty}$, or

$$k_i \sin\theta_i = k_t \sin\theta_t , \qquad (5.23)$$

which in terms of wavelength is

$$\frac{2\pi n_i}{\lambda_0}\sin\theta_i = \frac{2\pi n_t}{\lambda_0}\sin\theta_t. \tag{5.24}$$

Cancelling like terms gives us Snell's Law

$$n_i \sin\theta_i = n_t \sin\theta_t. \tag{5.25}$$

So the "laws" of reflection and refraction are derived directly from Maxwell's equations. We can use the boundary conditions given by Maxwell's equations to further characterize the reflected and transmitted fields in Fig. 5.3.

Now we simplify Eq. (5.19) using Eq. (5.20), and introduce reflection and transmission coefficients into Eq. (5.19). Considering the field orientations, the electric field boundary condition gives

$$E_{i0}\cos(\theta_i) + rE_{i0}\cos(\theta_r) = tE_{i0}\cos(\theta_t). \tag{5.26}$$

As in the case of normal incidence we will need another boundary condition in order to get separate expressions for r and t; on to the magnetic field boundary condition.

Writing an equation for the magnetic field boundary condition is simpler, since the magnetic fields in all cases in Fig. 5.3 are in the plane of the surface. The incident magnetic amplitude is the incident electric amplitude divided by the velocity of the wave, so having equal tangential magnetic fields on each side of the surface translates to

$$\frac{E_{i0}}{v_1} - r\frac{E_{i0}}{v_1} = t\frac{E_{i0}}{v_2}. \tag{5.27}$$

Note this equation is the same as our normal incidence analysis [Eq. (5.12)], since the magnetic fields for both are in the plane of the surface.

Now we can combine Eqs. (5.26) and (5.27) to obtain reflection and transmission coefficients.[13] By setting the incident to the reflected angle and using $v_1/v_2 = n_2/n_1$, Eq. (5.26) and Eq. (5.27) give

$$t = \frac{2n_1 \cos(\theta_i)}{n_2 \cos(\theta_i) + n_1 \cos(\theta_t)} \quad \text{and} \tag{5.28}$$

---

[13] You can derive equations (5.28) and (5.29) by algebraic manipulation or use the matrix theory you learned in linear algebra. If you have doubts about the latter ask your instructor.

$$r = \frac{n_1 \cos(\theta_t) - n_2 \cos(\theta_i)}{n_2 \cos(\theta_i) + n_1 \cos(\theta_t)}. \qquad (5.29)$$

You will note that these results agree with Eq. (5.14) and Eq. (5.15) when $\theta_i = 0$, i.e. normal incidence. The t and r expressed by Eqs. (5.28) and (5.29) are known as Fresnel transmission and reflection coefficients. In this case we have found the coefficients for light polarized parallel to the plane of reflection. The coefficients for light perpendicular to the plane of reflection will be different (Exercise 5.5). The reflection coefficients for both parallel and perpendicular polarizations are plotted in Fig. 5.4.

An important result comes out of a close examination of Eq. (5.29) Note r can be zero if

$$\frac{n_2}{n_1} \cos(\theta_i) = \cos(\theta_t); \qquad (5.30)$$

the reflection disappears for this condition. The incident angle at which this occurs is known as the Brewster angle, θ_B. By combining this Brewster condition [i.e. Eq. (5.30)] with Snell's law we find that

$$n^2 \cos^2(\theta_B) + \frac{1}{n^2} \sin^2(\theta_B) = 1, \qquad (5.31)$$

where $n = n_2/n_1$. The "physical" solution to Eq. (5.31) is $\theta_B = \tan^{-1}(n)$, see Exercise 5.4. This angle is known as Brewster's angle. Brewster's angle is important because of what it enables, e.g. the generation of polarized light from unpolarized light (e.g. light from the sun), and the generation of polarized light from lasers. Fig. 5.4 shows the intensity coefficient of reflection $R = r^*r$ as a function of the incident angle for two perpendicular polarizations. For the parallel polarization Brewster's angle is quite distinct. For the perpendicular polarization the reflected intensity is never zero. This disparity enables the use of a simple piece of glass for polarizing light. It also provides a tool for determining whether a laser is polarized (see Experimental Problem #2). Beyond this the Brewster equation, $\tan(\theta_B) = n$, allows one to determine the refractive index of a dielectric surface.

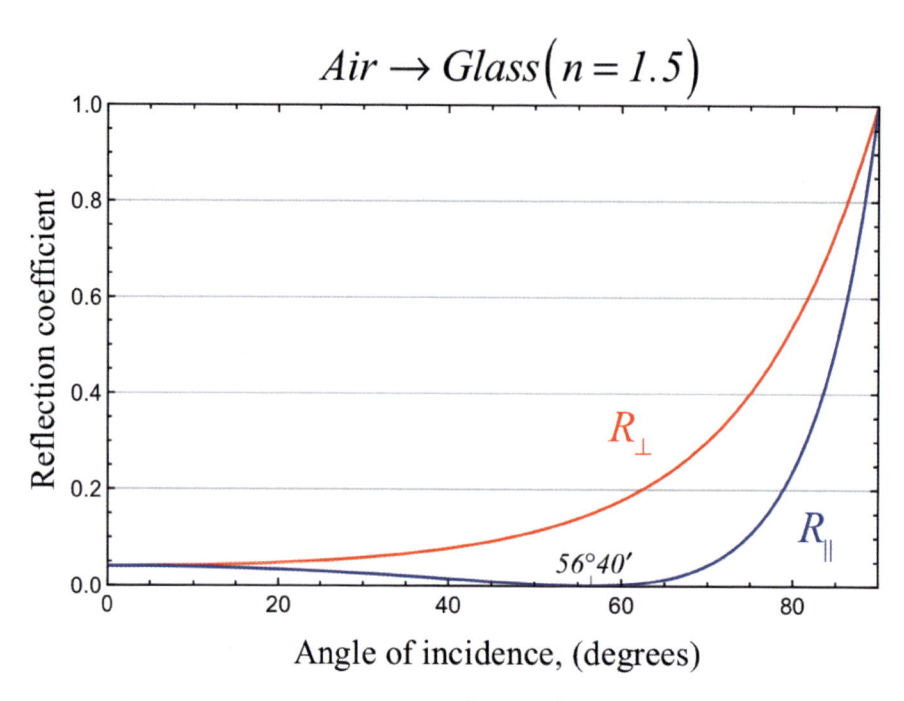

*Figure 5.4 – Reflection from a glass surface (**n~1.5**). When the field is parallel to the plane of reflection, we see an incident angle at which the reflection goes to zero, the Brewster angle, **θ$_B$**.*

5.3 – Total Internal Reflection and Evanescent Field

Something especially interesting happens when light approaches an interface from a region of higher refractive index than that on the transmitted side, Fig. 5.5. Let's analyze the components of the field in the figure to see what arises.

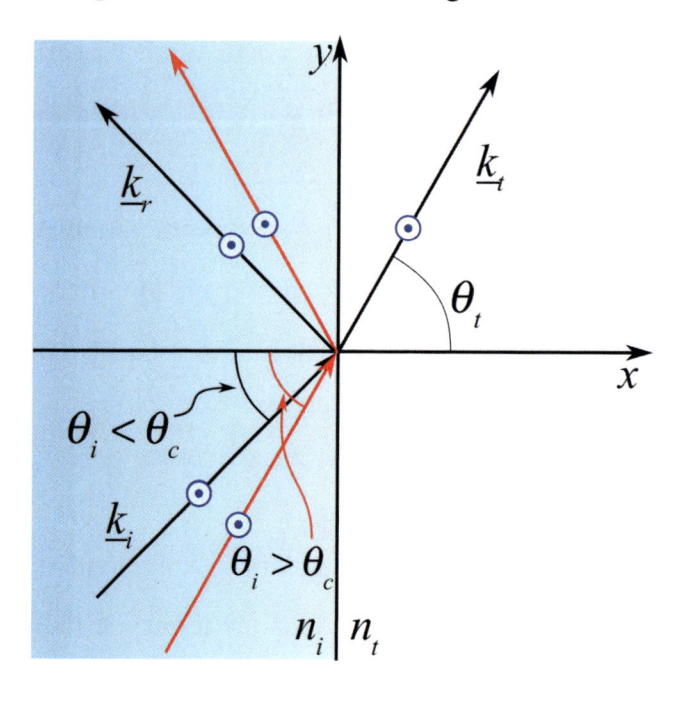

Figure 5.5 – Angles of incidence, reflection, and transmission at a boundary for which $n_i > n_t$.

All wave vectors lay in the *xy*-plane so the squares of the incident and transmitted wave vectors are

$$k_i^2 = k_{ix}^2 + k_{iy}^2 \text{ and} \tag{5.32}$$

$$k_t^2 = k_{tx}^2 + k_{ty}^2. \tag{5.33}$$

As in Eq. (5.20); $k_{iy} = k_{ty}$. We are interested in the *x*-component of the transmitted wave vector; from Eq. (5.33)

$$k_{tx} = \left(k_t^2 - k_{ty}^2\right)^{1/2} = \left(k_t^2 - k_{iy}^2\right)^{1/2} = \left(k_t^2 - k_i^2 \sin^2\theta_i\right)^{1/2} = k_0\left(n_t^2 - n_i^2 \sin^2\theta_i\right)^{1/2}. \tag{5.34}$$

In the region $n_i^2 \sin^2\theta_i \geq n_t^2$, k_{tx} is zero or purely imaginary. A more detailed analysis would show that in this region there is no loss of radiation in the *x*-direction. The critical angle θ_c is the smallest incident angle for which the propagation component in the *x*-direction in the transmitted medium is zero;

$$\sin\theta_c = \frac{n_t}{n_i}. \tag{5.35}$$

At and beyond the critical angle light propagates along the boundary but its energy rejoins with the reflected light. As θ_i becomes greater than the critical angle k_{tx} becomes imaginary, making it more convenient to write it as

$$k_{tx} = ik_0\left(n_i^2 \sin^2\theta_i - n_t^2\right)^{1/2}. \tag{5.36}$$

To make the next steps a bit neater we will use $\eta = \left(n_i^2 \sin^2\theta_i - n_t^2\right)^{1/2}$. Replacing k_{tx} with $ik_0\eta$ allows us rewrite the electric field as

$$E_t(x,y,t) = E_0 e^{-k_0\eta x} e^{i(k_{ty}y - \omega t)}. \tag{5.37}$$

The corresponding intensity is proportional to absolute squared of the field

$$I \propto e^{-2k_0\eta x}. \tag{5.38}$$

Beyond the critical angle where η is real and positive, the intensity decays exponentially above the surface in Fig. 5.5. The characteristic length, or distance at which the light has decayed to 1/*e* of the original strength is $L = \frac{1}{2k_0\eta}$. An amazing property of reflection beyond the critical angle is that it is without loss.

In this way so called "total internal reflection" beats reflection from a metallized mirror for efficiency. This property can be used to bottle light over an extended period, a subject we will return to in our chapter on "optical resonators". The exponentially decaying field produced, known as the "evanescent field", draws no energy from the light reflected at the surface of the material unless absorbing molecules are placed in this field.

We can compute the length L of the evanescent intensity from a prism having a refractive index $n_i = 1.6$, illuminated with light with a free space wavelength $\lambda = 500$ nm at angle $\theta_i = 80°$, and having water on the evanescent side with refractive index $n_t = 1.33$. To start we recognize that $k_0 = 2\pi/\lambda_0$, so that

$$L = \frac{\lambda_0}{\left[4\pi\left(n_i^2 \sin^2\theta_i - n_t^2\right)^{1/2}\right]} = \frac{500 \ nm}{4\pi\left(1.6^2 \sin^2(80°) - 1.33^2\right)^{1/2}} \approx 37 \ nm. \quad (5.39)$$

It should be noted that this length is much smaller than the length scale over which light can be focused by a lens. This opens up a great many applications for evanescent excitation by light, such as Total Internal Reflection Fluorescence Microscopy (TIRFM), which we will discuss in Chapter 10.

5.4 – Light Interactions with Metal

So far we have been talking about dielectric interfaces (non-conductors), however metals also comprise an important part of optics. For example, mirrors use thin films of silver or aluminum on dielectrics. In what follows we investigate how to describe the optical properties of metals.

Metals are characterized by having free electrons, in contrast to the bound electrons in atoms and molecules. Interactions with the surrounding lattice may be characterized by a drag coefficient C_d. So the springs in our Lorentz atom are gone, and a simple dynamical equation for the electron is

$$m_e \frac{d^2 y_e}{dt^2} = -C_d \frac{dy_e}{dt} - |q_e| E_0 \cos(\omega t). \quad (5.40)$$

This equation is trimmed by dividing through by the mass of the electron and defining C_d/m_e by a relaxation rate γ_e;

$$\frac{d^2 y_e}{dt^2} = -\gamma_e \frac{dy_e}{dt} - \frac{|q_e|}{m_e} E_0 \cos(\omega t). \qquad (5.41)$$

Certainly you should be asking questions like (1) how does the field penetrate into the metal? (2) What are the interactions with the surrounding electrons, lattice, etc.? However for now we will only say that an electric field can penetrate a thin metal especially when the frequency is large, and all other influences have essentially been lumped into γ_e.

Once again we represent the $\cos(\omega t)$ as $e^{-i\omega t}$, and y_e as $y_{e0} e^{-i\omega t}$ with the understanding that only the real part of y_e is relevant. On this basis the complex amplitude y_{e0} is given by

$$y_{e0} = \frac{|q_e| E_0 / m_e}{\omega^2 + i\omega \gamma_e}. \qquad (5.42)$$

A particularly interesting aspect of this result is that with very little dissipation the amplitude of the motion of the electron is in phase with the field. Since the electron is expected to move opposite to the field when the field is first applied one might expect a 180 deg. phase shift. Clearly, Eq. (5.42) does not include the transient signal associated with turning on the field. However that signal fades over a short time $\sim 1/\gamma_e$. We are interested in the steady state oscillation, and the phase reflected in that oscillation is a consequence of inertia. In a metal we are interested in the collective oscillation of a statistical number of electrons characterized by a number density ρ_N. This electron plasma driven by the field will cause a time dependent Polarization density with amplitude $P_0 = -\rho_N |q_e| y_{e0}$. On this basis the "relative permittivity" of a free electron metal from Eq. (3.39) is

$$\varepsilon_r = \frac{\varepsilon}{\varepsilon_0} = 1 + \frac{P_0}{\varepsilon_0 E_0} = 1 - \frac{\omega_p^2}{\omega^2 + i\omega \gamma_e}, \qquad (5.43)$$

where ω_p, which is known as the plasma frequency is given by

$$\omega_p^2 = \frac{\rho_N |q_e|^2}{\varepsilon_0 m_e}. \qquad (5.44)$$

There are many practical consequences of Eq. (5.43). I'll start by talking about mirrors.

A good mirror has high reflectivity in the visible. We can understand how this is possible by computing the refractive index of a material obeying Eq. (5.43) at visible frequencies. To get a feeling for this we will ignore the relaxation rate (i.e. set $\gamma_e = 0$). From Eq. (5.43) we see that the permittivity is negative if we use a frequency below the plasma frequency. This means that the refractive index would be purely imaginary. If you compute $R = r^*r$ from Eq. (5.14) using $n_1 = 1$ and take n_2 as any imaginary number you will discover that R is 1. The reality is that there is a relatively small real part to n_2 at visible frequencies that will keep the reflectivity of a mirror from being perfect.

One can apply Eq. (5.43) to a gaseous plasma, such as the ionosphere. Here the plasma electron density ρ_N is considerably smaller than in a metal, leading to a plasma frequency ~25 MHz and allowing γ_e to be reasonably set to zero. The ionosphere is a thick cold layer of ionized gas extending from ~80 km to ~180 km above the surface of the Earth. It is generated by ionizing radiation from the Sun, and consequently its electron density distribution differs throughout the day. It is very much like a mirror for radio waves with frequencies below the plasma frequency (see Ex. 5.11). However radio waves with frequencies above the plasma frequency are transmitted and can communicate with satellites.

5.5 – Chapter Five Exercises

Exercise 5.1 – You look directly at the water in a lake and see your reflection. Compute the fraction of the light intensity directed toward the water that is reflected.

Exercise 5.2 – What is the phase of the electric field reflected from the water, relative to the incident field?

Exercise 5.3 – Show that the right hand side of Eq. (5.16) is the intensity of the incident beam, by treating the refractive indices n_1 and n_2 in Eq. (5.14) and Eq. (5.15) as real.

Exercise 5.4 – Derive Eq. (5.31) and show that its solution is $\tan(\theta_B) = n$.

Exercise 5.5 – Eq. (5.28) and (5.29) are the Fresnel transmission and reflection coefficients for light polarized parallel to the plane of reflection, Fig. 5.3. Determine the Fresnel reflection and transmission coefficients for light polarized perpendicular to the plane of reflection.

Exercise 5.6 – Compute the reflection coefficient $R = r^*r$ for an aluminum mirror at $\lambda_0 = 500$ nm in air. The complex refractive index of aluminum at 500 nm is $n = 0.81257 + 6.0481\,i$.

Exercise 5.7 – Find the transient solution to Eq. (5.41) if the system is at rest and at the origin at $t = 0$. You should be able to show that the electron initially moves downward (i.e. against the field), whereas at long times its motion is in phase with the field.

Exercise 5.8 - At what incident angle θ_i is the Evanescent length L least? What is the value for this least L at a wavelength of 650 nm for a prism interface with air if the prism has a refractive index of 1.65?

Exercise 5.9 – Find the evanescence length of red light $\lambda_0 = 650\,nm$ incident on a silica air interface at an angle of 80°. The refractive index of silica is about 1.5.

Exercise 5.10 – Using Eq. (5.29) show that at grazing incidence (i.e. $\theta_i = 90°$) the amplitude of the reflection coefficient $r = +1$ independent of whether n_2 is greater or less than n_1. Can you find a way to understand the lack of phase change for grazing reflection?

Exercise 5.11 – Total Internal Reflection Fluorescence Microscopy (TIRFM) uses evanescent fields to excite fluorescent molecules in a sample as shown below. The fluoresced light is then imaged by an objective lens, as shown in the figure below. The use of the evanescent field has the advantage of increasing axial resolution by exciting only a thin layer of the sample.

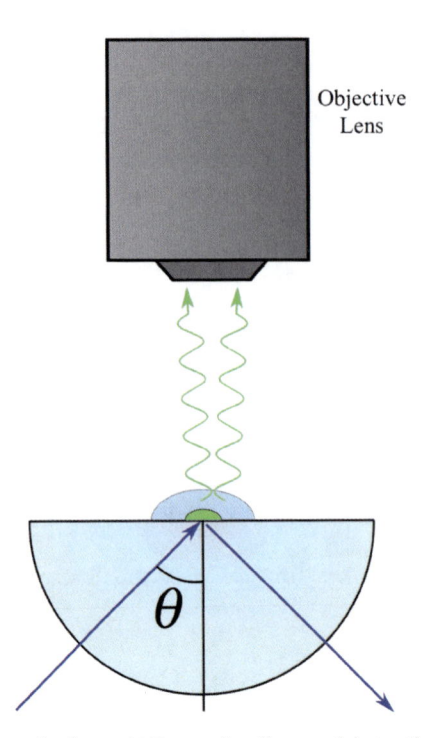

A laser beam with wavelength $\lambda_0 = 458\ nm$ is directed into the semi-circular glass prism with refractive index $n_g = 1.6$, shown in the figure above, and internally reflected against the flat face at an angle of incidence $\theta = 60°$. A water droplet above the prism with refractive index $n_w = 4/3$ contains fluorescent dye molecules. Calculate the thickness of the water droplet that is made to fluoresce assuming that this thickness corresponds to the height from the interface for the intensity to drop off by a factor of e (i.e. $e = 2.718$).

Exercise 5.12 – An electromagnetic wave (EM wave) is directed vertically upward from the Earth's surface with a frequency of 5 MHz. It interacts with the ionosphere layer 200 km above the Earth's surface where the electron relaxation rate γ_e can be neglected. The electron number density for this plasma layer gives it a plasma frequency 20 MHz.

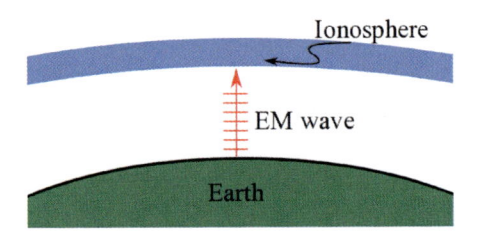

(a) If the ionosphere layer is treated as a free electron metal, what is the frequency dependence of its refractive index? **Hint**: refractive index $n = \sqrt{\varepsilon/\varepsilon_0}$.

(b) Does the oscillating electric field enter the plasma? If so, how does this field vary with penetration depth for the incident 5 MHz EM wave?

(c) If 10 W of power is incident on the layer, how much power is reflected back to the Earth?

Experiment #6 – Laser Polarization & Refractive Index Measurement using Brewster's law

Examine whether your laser is polarized by rotating the laser about its beam axis while observing the reflection from a surface. To get started first find the square plastic piece in your kit that is 20mm on a side. Next attach this piece to the black rotor plug by using the provided binder clip (as shown on rt.). Now plug this assembly in the center of the bearing in the baseboard. With the laser on, direct its beam along the line on the base plate, and collimate the light to produce a good focused beam by turning the lens at the front of the laser. Now rotate the assembled piece until the laser light hits its surface at an incident angle of ~60° from the normal. Examine the intensity of the reflected light while rotating your laser about its beam axis. If the laser is polarized you should observe a modulation in the reflected light intensity as you rotate the laser. From the condition for which the reflection is minimum in intensity estimate the polarization direction of the laser (hint: the minimum should occur when the polarization of the laser is parallel to the surface of the base plate). Does this polarization direction agree with what you obtained from the Rayleigh scattering experiment ? Mark the polarization direction of the laser on its body, and leave it in this polarization state relative to the baseboard. You will find that your laser is not fully polarized; there will be a slight amount of light polarized perpendicular to the baseboard. Before measuring the Brewster angle add a polarizer between the laser and rotation assembly so that its polarization axis is parallel to the base plate, in order to purify the parallel polarization as seen by the dielectric reflector. Now you are ready to measure the Brewster angle of the reflecting surface. Measure the Brewster angle by rotating the dielectric plate until you get a close to zero reflected intensity reflection. Calculate the refractive index of the cutout from Eq. (4.1). Based on your refractive index measurement is the cutout PMMA or Polystyrene (PS)? Info: at 652nm, $n_{PMMA} = 1.488, n_{PS} = 1.586$.[14]

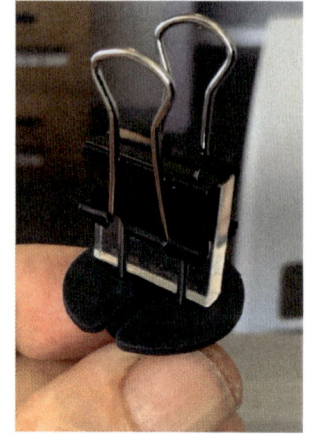

---

[14] Data obtained from internet at refractiveindex.info

My Pocket Optics Kit set-up for this experiment is shown below. In the image the reflecting suface is set to an incident angle of just under 50 deg. Note the reflection on the blue screen (upper rt.).

Scan for a video demo of this experiment

Experiment#6 Setup

Pedagogy – Sections 4.3 and 5.2 – Brewster angle

141

Chapter Six – Optical Resonators

6.1 – Resonators as tuners: From macro-AM tuners to Plasmonic PetaHz-nano-tuners

Resonators in electrical engineering are often used as tuners. That's how I first became aware of them. I will start this discussion by talking about something I built as a kid. It's the thing that first made me wonder about resonators. Resonators contain oscillating electric and magnetic fields over an extended time. By that I mean a time much longer than the time for light to move across the device. After first describing my inception into the world of macro-resonators, I will describe resonators that are nanoscopic.

I built my first resonator before I understood its theory. I asked my father about a radio receiver and he bought me a kit of parts to build the circuit in Fig. 6.1. It is the electrical schematic for a crystal radio.

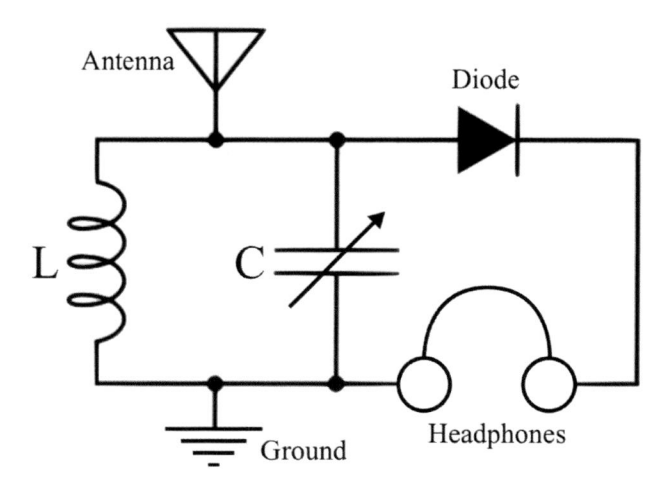

Figure 6.1 – Crystal radio.

Note that the crystal radio has no battery. The energy to power it comes from electromagnetic energy in the air. It is an AM radio receiver that tunes into stations by the use of the parallel inductance L and capacitance C. An inductor and capacitor in parallel is a resonant circuit with an angular resonance frequency given by

$$\omega_r^2 = \frac{1}{LC}. \tag{6.1}$$

This LC "tuner" is stimulated by an antenna and selects stations using a variable capacitor. At resonance energy builds up in the LC combination. The AM radio tunes into stations between 500 kHz and 1500 kHz. Is it possible to extend the use of discrete L and C components for a resonator to the frequency of visible light (i.e. around 10^{15} Hz)?

The short answer is no. We will require a new approach, as store bought inductors and capacitors won't take us there; their values of L and C are too large (smallest C is about 0.1 pF and smallest L is about 1 μH would make $f_r = 5 \times 10^8 \, Hz$). Surprisingly, the solution is simple, a metallic nano-spheriod (e.g. Gold or Silver).

Fig. 6.2 shows the scattering spectrum from a silver nanospheriod ~ 35 *nm* in size (blue, left), at the center of an image of a group of particles (right). The blue spectrum has a Lorentzian-like shape, consistent with a resonator. In effect, the particle excited along its long axis is a nano-filter even though there is no obvious LC-circuit. As analyte molecules bind to the spheroid the resonance spectrum shifts, indicating that the nanospheriod can be used as a nano-sensor. How is the resonance possible?

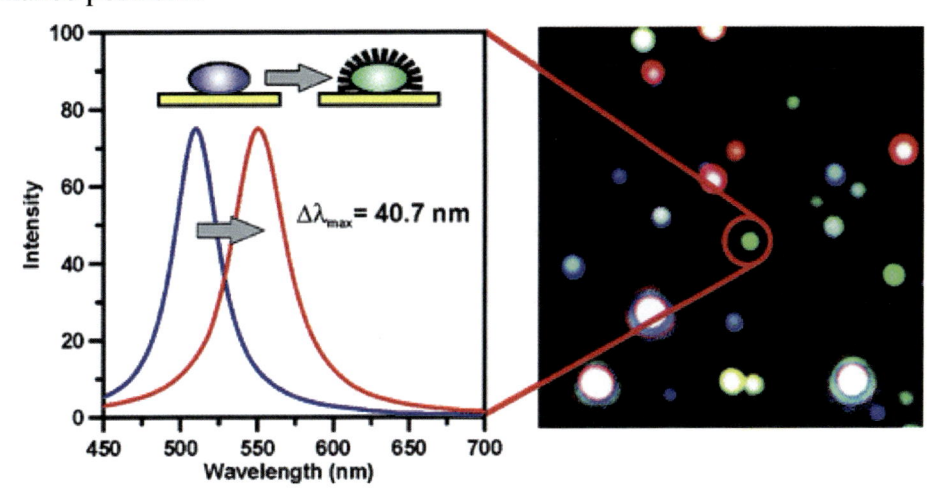

*Figure 6.2 – The scattering spectrum from a spheroidal silver nanoparticle as a function of the wavelength of the incident light. The electric field of the light bathing this particle is polarized along the long axis of the spheroid. (Reprinted with permission from A. MacFarland & R.P. Van Duyne, Nano Letters, **2003**, 3(8), pp 1057-1062. Copyright 2003 American Chemical Society.)*

The dielectric function of a metal in Eq. (5.43) clearly does not show any obvious resonant characteristics; its form is not Lorentzian. You might guess at this point that geometry plays a role. Indeed it does, with the most important aspect being confinement. The mobile electrons in a metallic nano-particle are confined by their geometrical constraints.

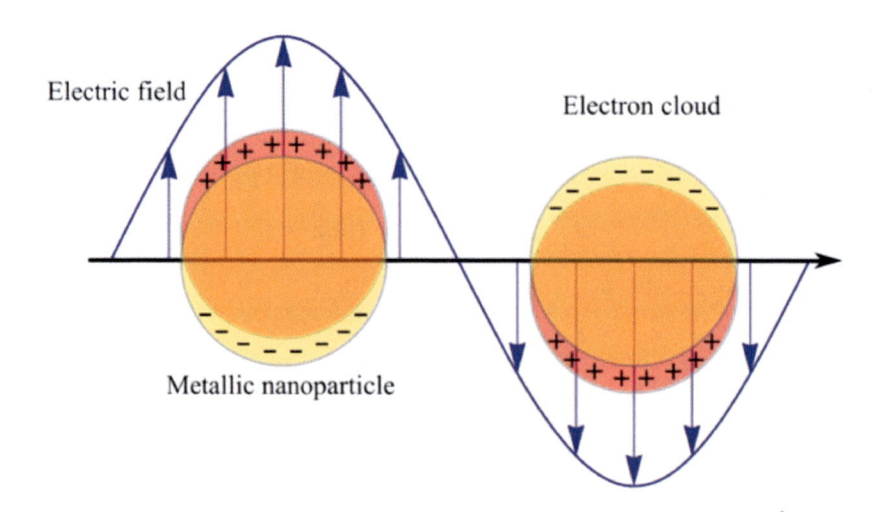

Figure 6.3 – A metallic nanoparticle driven by a plane wave.

Fig. 6.3 illustrates a neutral silver nanoparticle driven by an external wave at a frequency below the resonance frequency. When the field is upward the free electrons are driven downward, thereby uncovering positive lattice charge on the upper half of the nanoparticle. If now we suppose the field is turned off, then the electrons at the bottom will rush back under the pull of the positive charge on the top. Since the electrons have inertia, their center of negative charge will move above the center of positive charge, and the electrons will be attracted downward. The system is an oscillator, which can be driven into resonance. To analytically understand this resonance requires looking at the dynamical behavior of the electrons.

I will focus on the centroid of negative charge $-|Q_e|$. When the centroid moves upward from center, free of an external wave, it leads to an imbalanced situation in which there is more positive charge below it to pull it back down. It is just like a "gravity train" in mechanics, except that gravitational forces are not at work here. Instead, an electric field develops that pulls on the centroid of negative charge. That field is proportional to the displacement y_e and to the density of positive charge ρ_+,[15] and is given by Gauss's law to be $E = \rho_+ y_e/(3\varepsilon_0)$ (Ex. 6.5). This leads to a restoring force

$$F = \frac{-|Q_e|\rho_+ y_e}{3\varepsilon_0}, \qquad (6.2)$$

that is linear with displacement, just like the force from a "spring" having spring constant $\frac{|Q_e|\rho_+}{3\varepsilon_0}$. Given the electron mass and this spring constant, oscillation will ensue, with frequency

$$\omega_{LSP}^2 = \frac{|Q_e|\rho_+}{3\varepsilon_0 M_e}, \qquad (6.3)$$

where M_e is the mass of all the free electrons.[16] But the charge to mass ratio of all of the electrons is the same as the elemental charge divided by the electron's mass so

$$\omega_{LSP}^2 = \frac{|q_e|\rho_+}{3\varepsilon_0 m_e}. \qquad (6.4)$$

In addition the positive charge density is just the elemental charge times the number density of free electrons $\rho_+ = |q_e|\rho_N$. With all this in hand, $\omega_{LSP}^2 = \omega_p^2/3$, where ω_p is the so-called plasma frequency from Eq. (5.44); $\omega_p^2 = \rho_N |q_e|^2/(\varepsilon_0 m_e)$. With these simplifications the dynamical equation for the centroid position is

$$\frac{d^2 y_e}{dt^2} = -\frac{\omega_p^2}{3} y_e - \gamma_p \frac{dy_e}{dt} - \frac{|Q_e|}{M_e} E_0 \cos(\omega t). \qquad (6.5)$$

Eq. (6.5) is a typical equation of a driven harmonic oscillator. Resonance is assured so long as the relaxation rate γ_p is not too large. Although for most metals the plasma frequency ω_p is in the Ultra-Violet, the resonant frequency in the sphere is considerably lower by a factor of $\sqrt{3}$. The characteristic lifetime of the oscillation is $1/\gamma_p$. The subject built around confining energy at the

---

[15] This charge density is composed of the net positive charge of each lattice atom.
[16] The LSP subscript stands for local surface plasmon.

frequency of visible light by using an electron "plasma" in a metal or semiconductor is known as "plasmonics". To see the resonance more clearly one can use the dynamical response to get the frequency dependence of the dipole moment.

In the usual way we represent $\cos(\omega t)$ as $e^{-i\omega t}$, and y_e as $y_{e0} e^{-i\omega t}$, from which y_{e0} is found to be

$$y_{e0} = \frac{|Q_e|}{M_e} \frac{E_0}{\omega^2 - \omega_p^2/3 + i\omega\gamma_p}. \qquad (6.6)$$

The resonance is clearly seen in Eq. (6.6). The dipole amplitude is obtained from this expression by multiplying by $-|Q_e|$ with the result

$$\mu_{p0} = -|Q_e| y_{e0} = -\frac{|Q_e|^2}{M_e} \frac{E_0}{\omega^2 - \omega_p^2/3 + i\omega\gamma_p}. \qquad (6.7)$$

This provides an equation for the polarizability of the form

$$\alpha_{LSP} = \frac{|Q_e|^2}{M_e} \frac{1}{\omega_p^2/3 - \omega^2 - i\omega\gamma_p}. \qquad (6.8)$$

An alternate way to write Eq. (6.8) shows that polarizability is proportional to the volume of the particle V_p (see Ex. 6.2),

$$\alpha_{LSP} = \varepsilon_0 V_p \frac{\omega_p^2}{\omega_p^2/3 - \omega^2 - i\gamma\omega} \qquad (6.9)$$

There is a host of information that can be obtained from Eq. (6.9) with respect to the confinement of optical energy. For plasmonics, we localized the energy in a nanoscopic particle; next we will trap propagating photons.

6.2 – The Fabry-Perot Etalon

What you see in Fig. 6.4 is light with amplitude E_0 reflecting and being transmitted into a piece of glass having a thickness d.

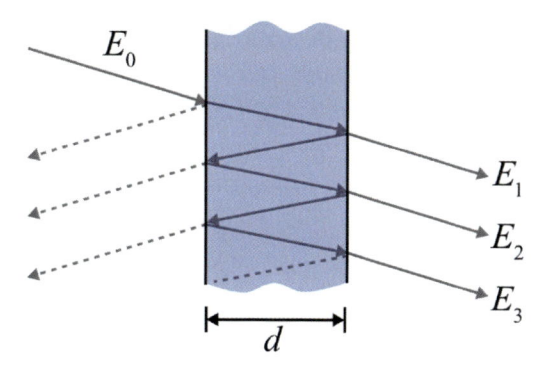

Figure 6.4 – Light interacting with an etalon.

Let's suppose that the polarization of the light is perpendicular to the figure. If the sides of the glass are polished to be very parallel and smooth, this device is known as an Etalon. It is a special case of what is commonly known as a Fabry-Perot resonator. By now you know that this means a device that can store energy over a limited time. To see why photon energy can build up in such a device it is necessary to analyze how the light incident on the Etalon is reflected by it, and transmitted through it.

I have not drawn all possible rays, notably I have left out some of the rays that should be emerging from the left after being transmitted into the Etalon. I am more concerned with what is contained within and transmitted through. In addition even though the light is shown reflecting at an arbitrary angle, I am interested in directing the light perpendicular to the glass. An incident angle other than zero degrees is shown in order to distinguish the rays. I will use the reflection and transmission coefficients that we developed in Section 5.1. Let's commence. The 1st ray entering the glass comes in with a field tE_0 and picks up a phase $e^{i\delta}$ in propagating to the right side of the glass, where $\delta = n_2 k_0 d$ with n_2 being the refractive index of the glass and k_0 the propagation constant in free space. Energy transmitted at this point is described by a wave $E_1 = t'e^{i\delta}tE_0$, where t' is the transmission coefficient from glass back into air. Only a fraction of the transmission gets out this way. The part of the wave not transmitted the first time must reflect from the inside surface, bounce against the left surface and transmit through the right hand side of the glass. Considering the added distance traveled in addition to the two added reflections the field of the second transmitted wave is

$E_2 = t'e^{i\delta}r'e^{i\delta}r'e^{i\delta}tE_0$, where r' is the reflection coefficient against the inside of the glass. You should see a pattern, but to be sure of that I will go one more time. The 3rd transmission $E_3 = t'e^{i\delta}r'e^{i\delta}r'e^{i\delta}r'e^{i\delta}r'e^{i\delta}tE_0$. All that we know thus far is $E_t = E_1 + E_2 + E_3 + ...$ or

$$E_t = t'e^{i\delta}tE_0 + t'e^{i\delta}r'e^{i\delta}r'e^{i\delta}tE_0 + t'e^{i\delta}r'e^{i\delta}r'e^{i\delta}r'e^{i\delta}r'e^{i\delta}tE_0 + ... \quad (6.10)$$

Eq. (6.10) can be condensed

$$E_t = t'e^{i\delta}tE_0\left(1 + (r')^2 e^{i2\delta} + (r')^4 e^{i4\delta} + ...\right). \quad (6.11)$$

We note that $(r')^2 = r^2$ since as we had shown earlier $r' = -r$; see Eq. (5.14). By defining $x = r^2 e^{i2\delta}$, the trend becomes clear;

$$\left(1 + r^2 e^{i2\delta} + r^4 e^{i4\delta} + ...\right) = \left(1 + x + x^2 + ...\right), \quad (6.12)$$

and if we think about N terms

$$\left(1 + r^2 e^{i2\delta} + r^4 e^{i4\delta} + ... + r^{N-1} e^{i(N-1)\delta}\right) = \left(1 + x + x^2 + ... + x^{N-1}\right). \quad (6.13)$$

This series can be identified as a power series (see Appendix A.3) of the form

$$\sum_{n=0}^{N-1} x^n = \frac{1-x^N}{1-x}, \quad (6.14)$$

for $x < 1$. Since x is less than 1 (the reflection coefficient must be less than 1), for an infinite number of terms

$$\lim_{N \to \infty} \frac{1-x^N}{1-x} = \frac{1}{1-x}. \quad (6.15)$$

By combining this sum with Eq. (6.11) the total transmitted field is found to be

$$E_t = t'te^{i\delta}E_0\left[\frac{1}{1-r^2 e^{i2\delta}}\right] \quad (6.16)$$

Just one more step: After using the transmission and reflection relations Eq. (5.14) and Eq. (5.15) we can see that $t't = 1 - r^2$ from which the ratio of the transmitted to incident intensity is

$$\frac{I_t}{I_0} = \left|\frac{E_t}{E_0}\right|^2 = \frac{(1-r^2)^2}{|1-r^2 e^{i2\delta}|^2}. \quad (6.17)$$

Eq. (6.17) is an amazing result. Although both sides of the Etalon reflect light there are phases 2δ at which all of the incident intensity is transmitted. These values are $2\delta_m = 2\pi m$ where $m = 1, 2, 3, \ldots$. If you think carefully about this result you will realize that each of these are resonances of the electromagnetic field within the Etalon (Fig. 6.5). They are similar to particle-in-a-box modes in Quantum Mechanics, or the string modes on a violin.

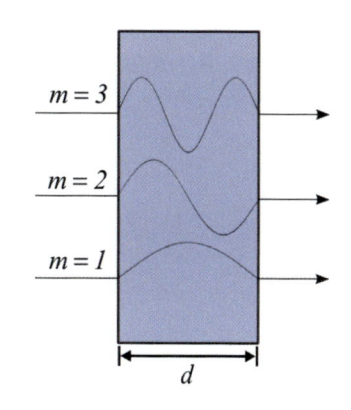

Figure 6.5 – First three mode shapes.

The spectrum of a typical resonance is shown in Fig. 6.6.

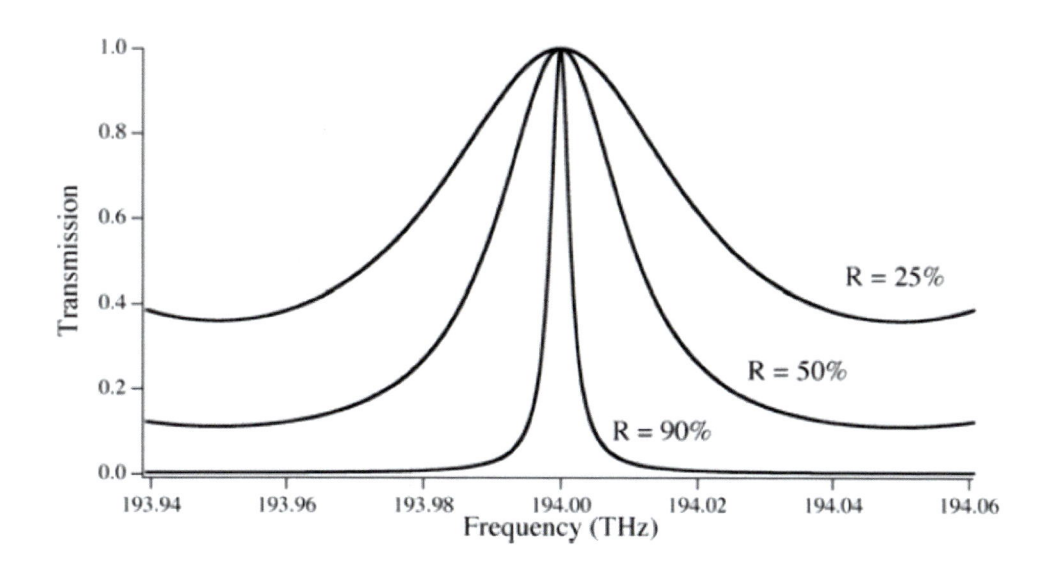

Figure 6.6 – Resonance of an etalon showing the effect on reflectance $R = r^*r$. For $\text{Im}[r] = 0$, $R = r^2$.

The picture in Fig. 6.4 does not go back to first principles; the direct use of Maxwell's equations. By combining Faraday's Law with the Maxwell-Ampere equation (with some help from Gauss's Law), we were able to derive the wave equation in Chapter One. We would like to derive the solution for the Fabry-Perot etalon along the same lines. Sending waves to a slab of glass results in reflection from the nearest surface, transmission into the slab, and reflection from as well as transmission through the second surface. The solution ultimately involves solving the wave equation in each of the regions. The expectation is that the answer for the transmission will agree with what we have obtained before Eq. (6.17).

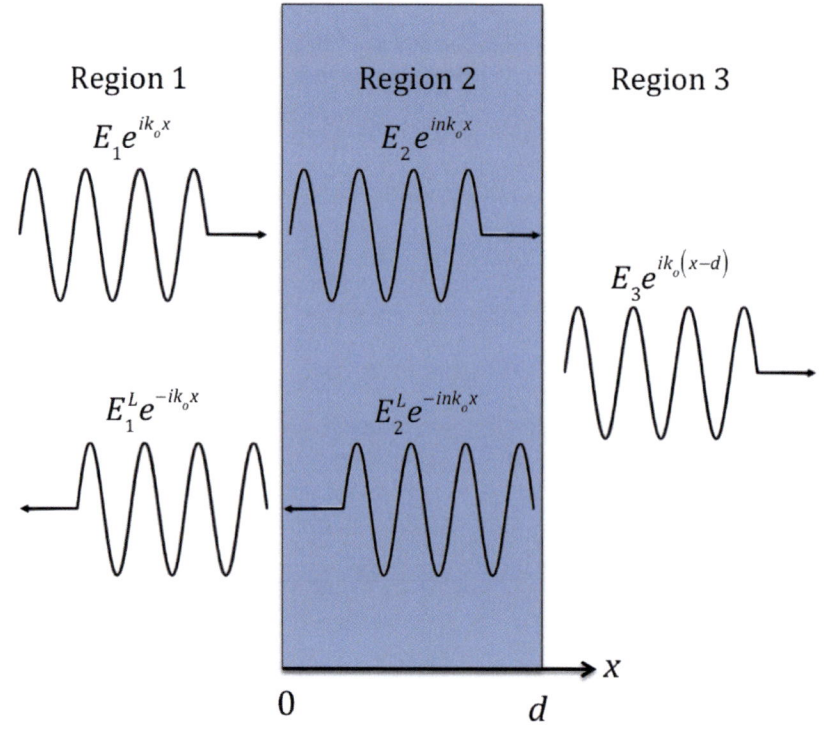

Figure 6.7 - Wave transmission through a dielectric slab.

Fig. 6.7 shows the basic setup. There are three regions – (1) to the left of the surface at $x<0$ for which the subscript 1 is used on the fields; (2) the region inside the slab from $0 \leq x \leq d$; and (3) the region to the right of the slab for which $x>d$. Note that the superscript L denotes a wave moving to the left. All waves are polarized along y and they are intended to be written as $E_2 e^{ink_o x}$ for the wave moving to the right, and $E_2^L e^{-ink_o x}$ for the wave moving toward the left.

There are two boundary conditions, and two interfaces to apply them to, resulting in four equations. The first boundary condition is on the electric field, and the other is on the magnetic field. The boundary condition on the electric field, a product of Faraday's equation, states that the total electric field on the left side of a boundary should equal the total electric field on the right side of a boundary. For the first boundary the electric field on the left is $E_1 + E_1^L$ and that on the right is $E_2 + E_2^L$, so our first equation is

$$E_1 + E_1^L = E_2 + E_2^L. \qquad (6.18).$$

The temporal derivative of the magnetic field is related to the spatial derivative of the electric field through Faraday's Law,

$$\frac{\partial B_z}{\partial t} = -\frac{\partial E_y}{\partial x}. \qquad (6.19)$$

Given the harmonic phase $e^{-i\omega t}$, the magnetic field for the incident wave and reflected wave at the boundary are

$$B_1 = \frac{k_0}{\omega} E_1 \text{ and} \qquad (6.20)$$

$$B_1^L = -\frac{k_0}{\omega} E_1^L, \qquad (6.21)$$

respectively. It should be noted that all electric fields are in the y direction and all magnetic fields are in the z direction. The magnetic fields on the right-hand side of the boundary are simply changed in as much as the characteristic propagation constant is multiplied by the refractive index of the slab,

$$B_2 = \frac{nk_0}{\omega} E_2 \text{ and} \qquad (6.22)$$

$$B_2^L = -\frac{nk_0}{\omega} E_2^L. \qquad (6.23)$$

The preservation of the total magnetic field on either side of the boundary gives

$$E_1 - E_1^L = nE_2 - nE_2^L. \qquad (6.24)$$

Eq. (6.18) and Eq. (6.24) have more than two unknowns, so more conditions will be needed to determine the strength of the reflected field in relation to the incident field. This should not be a surprise since the second surface plays an important role (i.e. it reflects energy back to the first surface). Although spatial phases are missing from Eq. (6.18) and Eq. (6.24) the second surface is not at the origin, so phases will have to be preserved in applying the boundary conditions at the second surface relative to the first surface, for the fields in the slab. The electric field boundary condition for the $x = d$ surface gives

$$E_2 e^{ink_0 d} + E_2^L e^{-ink_0 d} = E_3, \qquad (6.25)$$

and the magnetic field boundary condition at this surface is

$$nE_2 e^{ink_0 d} - nE_2^L e^{-ink_0 d} = E_3. \qquad (6.26)$$

The approach to solving for the transmission coefficient for the entire slab from Eq. (6.18), Eq. (6.24), Eq. (6.25), and Eq. (6.26) is to eliminate the fields in the slab first, by solving Eq. (6.25) and Eq. (6.26) in terms of E_3,

$$E_2 = \frac{n+1}{2n} E_3 e^{-ink_0 d} \text{ and} \tag{6.27}$$

$$E_2^L = \frac{n-1}{2n} E_3 e^{ink_0 d}, \tag{6.28}$$

and substituting these equations back into Eq. (6.18) and Eq. (6.24), with the results

$$E_1 + E_1^L = \frac{E_3}{2n}\left[(n+1)e^{-ink_0 d} + (n-1)e^{ink_0 d}\right] \text{ and} \tag{6.29}$$

$$E_1 - E_1^L = \frac{E_3}{2}\left[(n+1)e^{-ink_0 d} - (n-1)e^{ink_0 d}\right]. \tag{6.30}$$

Adding Eq. (6.29) to Eq. (6.30) and inverting the result yields

$$E_3 = \frac{E_1}{\cos(nk_0 d) - i\left(\dfrac{n^2+1}{2n}\right)\sin(nk_0 d)}. \tag{6.31}$$

From Eq. (6.31) the total intensity transmission coefficient is

$$T = \left|\frac{E_3}{E_1}\right|^2 = \frac{1}{\left|\cos(nk_0 d) - i\left(\dfrac{n^2+1}{2n}\right)\sin(nk_0 d)\right|^2}, \tag{6.32}$$

which simplifies to

$$T = \left|\frac{E_3}{E_1}\right|^2 = \frac{1}{1+\left(\dfrac{n^2-1}{2n}\right)^2 \sin^2(nk_0 d)}. \tag{6.33}$$

Now we see quite distinctly that the transmission coefficient reaches unity when $nk_0 d = m\pi$, where m is a positive integer. If the refractive index were very large, i.e. $n = 5$, the transmission would be as shown in Fig. 6.8.

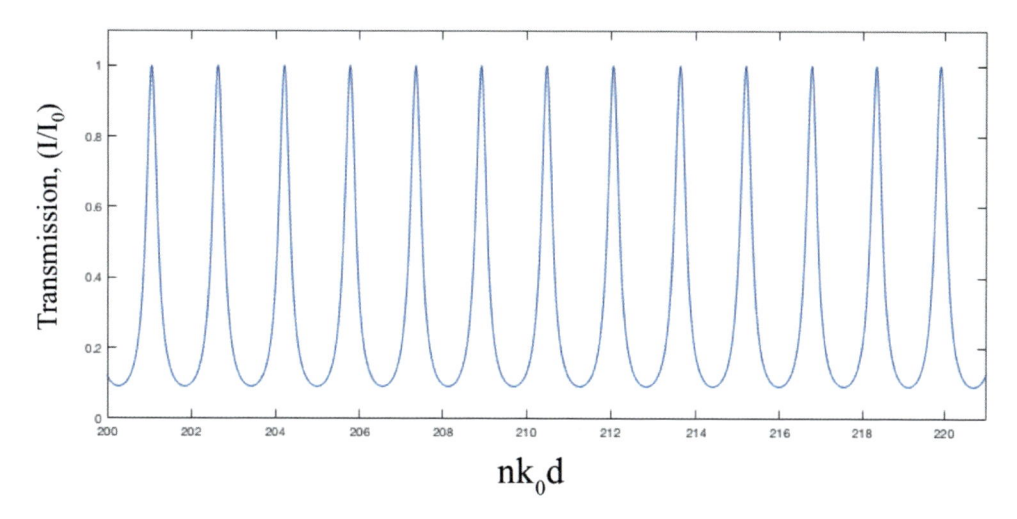

Figure 6.8 – The intensity transmitted through an etalon with refractive index n=5 vs. phase nk_0d, Eq. (6.33). At resonance the light is entirely transmitted even though there are "expected" losses from reflection.

Equations (6.17) and (6.33) are actually identical, a fact you are asked to prove in Ex. 6.8.

Before leaving the topic of the light transmission by a ultra-parallel slab of transparent material (Etalon), it is important to note that the intensity reflection coefficient R can approach 1 by depositing a metallic coating on the sides. In addition two partially transmiting mirrors with thier reflecting sides facing each other and separated by an air gap is commonly referred to as a Fabry-Perot Interferometer. If the airgap is controlled this configuration becomes a mechanically tunable optical filter known as a Fabry-Perot Spectrometer.

6.3 – Frequency and Time Domain as Conjugates

A question can be asked as to the sorts of measurements that lead to information concerning resonators. The Lorentzian shapes we saw in Fig. 6.2 and Fig. 6.6 are called frequency domain (FD) measurements. For example, you continuously change the color of light incident on a resonator, such as a metallic nanoparticle, while recording the scattering spectrum from the particle (Fig. 6.9).

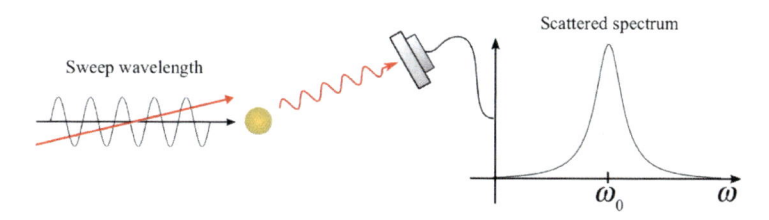

Figure 6.9 – Frequency-domain (FD) measurement of scatter from a resonant particle.

153

The conjugate to this measurement is one in the time domain (TD). A time domain measurement could involve directing a very short pulse of light on a particle and recording the light scattering. If you could resolve the actual field picked up by a detector you would identify an exponentially decaying oscillation occurring after the pulse (Fig. 6.10).

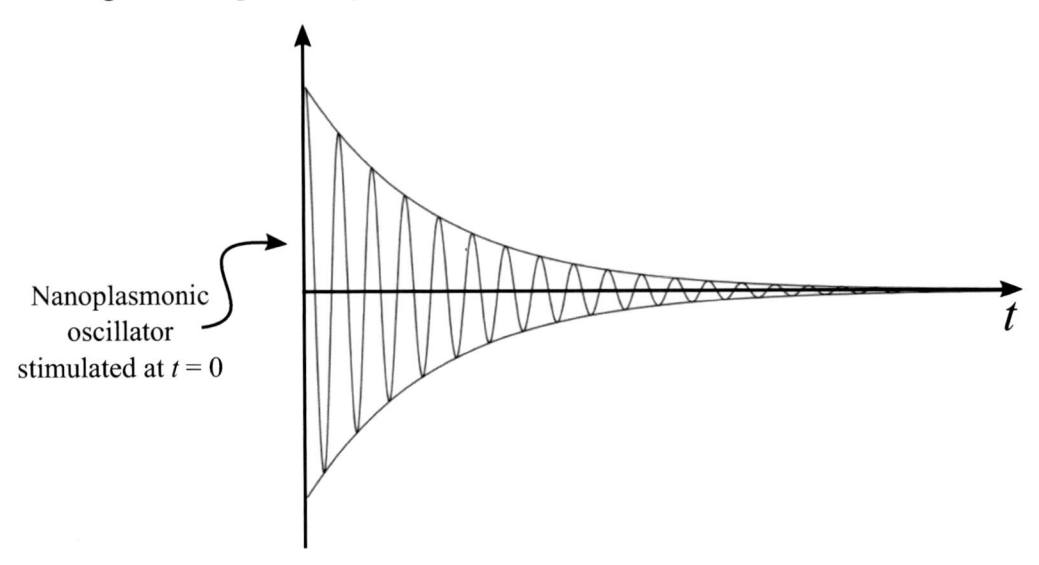

Figure 6.10 – Time-domain (TD) measurement of scattering from a resonant particle.

The amazing thing is that the same information can be extracted from either experimental approach because these are conjugate domains of measurement; a Fourier transform relates them. We have witnessed conjugates in diffraction theory in Chapter 2. There it was a spatial coordinate y and spatial frequency k_y. Now that we are talking about time and frequency domain, the conjugates are t and ω. The relationship between these conjugates is

$$H(\omega) = \int_{-\infty}^{\infty} f(t) \cdot e^{-i\omega t} dt \qquad (6.34)$$

The scattering intensity is

$$I = |H(\omega)|^2 \qquad (6.35)$$

To see how this works we represent the damped harmonic time response in Fig. 6.10 as

$$f(t) = \begin{cases} Ae^{-\frac{\gamma}{2}t} \cos(\omega_0 t), & t > 0 \\ 0, & t < 0 \end{cases} \qquad (6.36)$$

With this time behavior

$$H(\omega) = \frac{A}{2}\int_0^\infty \left[e^{i(\omega_0-\omega+i\gamma/2)t} + e^{i(-\omega_0-\omega+i\gamma/2)t} \right] dt$$
$$= \frac{-A/2}{i(\omega_0-\omega+i\gamma/2)} - \frac{A/2}{i(-\omega_0-\omega+i\gamma/2)}$$
(6.37)

Clearly the resonance at $\text{Re}[\omega] = -\omega_0$ is not physical, so we will just use the first term, from which the power spectrum is

$$|H(\omega)|^2 = \frac{(A/2)^2}{(\omega-\omega_0)^2+(\gamma/2)^2},$$
(6.38)

which has a Lorentzian form similar to the scattering spectrum illustrated in Fig. 6.9. What is interesting about the Lorentzian form is that the width at half maximum $(\delta\omega)_{1/2}$ is simply γ. But this is also the decay rate of the oscillator's energy $|f(t)|^2$ following a very short excitation pulse. If I characterize this energy decay by a lifetime $\tau = 1/\gamma$, and identify γ as the line width $(\delta\omega)_{1/2}$ measured in the frequency domain then

$$\tau(\delta\omega)_{1/2} = 1.$$
(6.39)

This is a very powerful relationship that says that the shorter (longer) the lifetime, the wider (narrower) the Lorentzian line shape. In other words the line width of the Lorentzian can be obtained from the decay time; there is equivalent information in both types of experiments. Eq. (6.39) can be very useful when taking a photon point of view in analyzing problems.

As an example of such a problem, let's suppose we would like to estimate the line width of a cavity composed of 2 mirrors separated by distance d. One is almost perfect, $R_1 = 1$, and the other which is imperfect; has $R_2 < 1$. Some laser cavities are like this. An easy way to estimate the line width would be to first write down an equation for the photon lifetime in the cavity and then use Eq. (6.39) to get the line width. Now imagine N photons bouncing between the mirrors. When any of these photons strike the mirror on the left they are reflected, however when they strike the mirror on the right there is a small probability of transmission.

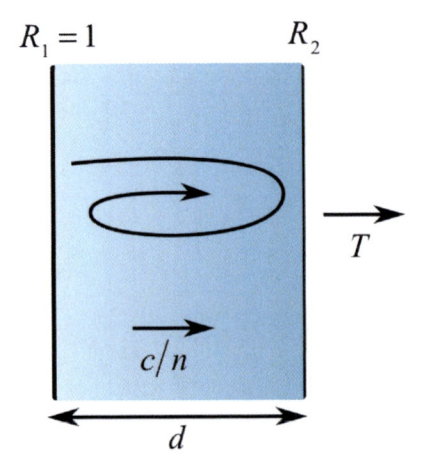

Figure 6.11 – A cavity for which the left mirror is perfect and the right mirror is leaky.

What we would like to know is the rate of loss of photons, because this depletes the photons in the cavity. Clearly this involves two things: the rate at which photons reach mirror 2, which we will call f_p and the probability P_T that upon reaching the mirror the photon will be transmitted (i.e. lost). The first part is easily answered - the time between hits on mirror 2 is the round trip time, $2dn/c$. Consequently the rate of hits on this leaky mirror is the inverse, $f_p = c/(2nd)$.

The probability of transmitting is just $1 - R_2$, so the rate of loss of photons

$$\frac{dN}{dt} = -f_p(1-R_2)N = -\frac{c(1-R_2)N}{2nd}, \tag{6.40}$$

where N is the number of photons in the cavity. From the structure of this equation it is apparent that there will be an exponential decay of the photon number with time,

$$N(t) = N_0 e^{-\frac{t}{\tau}} = N_0 e^{-\frac{(1-R_2)c}{2nd}t}. \tag{6.41}$$

This identifies

$$\tau = \frac{2nd}{(1-R_2)c}. \tag{6.42}$$

From Eq. (6.39) we estimate the line width of the cavity to be[17]

$$(\delta\omega)_{1/2} = \frac{(1-R_2)c}{2nd}. \tag{6.43}$$

As we discussed earlier in this sections there are some frequencies of light that will resonate in a cavity with two reflecting surfaces. The resonant modes of the cavity resemble the standing waves that appear on a vibrating string, see Fig. 6.12.

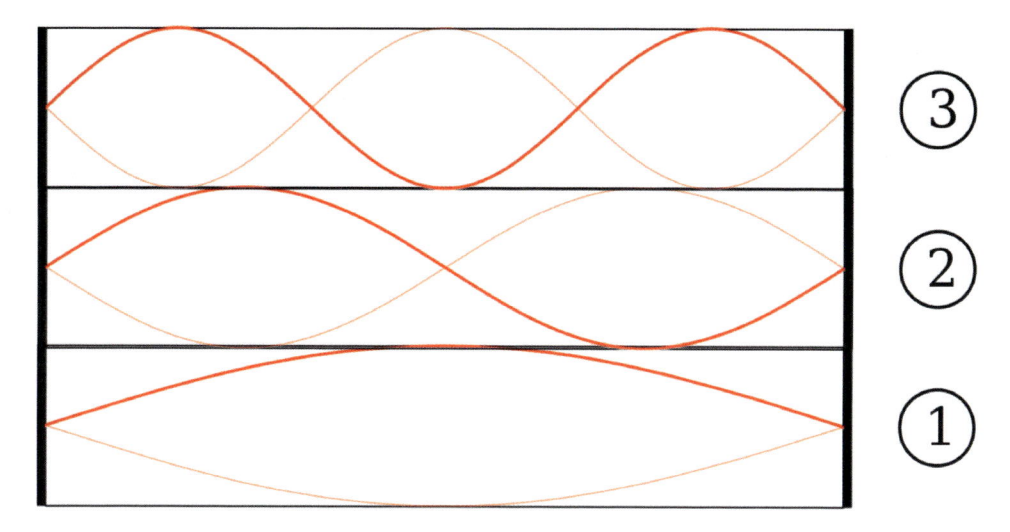

Figure 6.12 – The first three modes of a resonant cavity or string with two fixed ends.

In the resonant cavity these modes occur when $k_m d = m\pi$. Substituting $k_m = nk_{0,m}$ and $k_{0,m} = \omega_m/c$ gives the frequency of the mth mode ω_m the cavity;

$$\omega_m = \frac{\pi c}{nd} m. \tag{6.44}$$

It is common to characterize the longevity of light in a resonator by the quality factor Q. Q is the frequency of a mode divided by the line width, and is therefore proportional the photon lifetime τ in Eq. (6.39). By combining Eq. (6.43) and Eq. (6.44), Q is found to be

$$Q = \frac{\omega_m}{(\delta\omega)_{1/2}} = \frac{2\pi m}{(1-R_2)}. \tag{6.45}$$

---

[17] Our analytical approach of turning discrete bounces into a continuous differential equation is more realistic for a small fractional change in N per bounce. This is equivalent to having R_2 close to 1.

Note that as R approaches 1, Q approaches infinity; the resonator is sealed from losses.

We introduced a lot of new nomenclature in this section. Here is a brief review of the nomenclature for resonators:

- The *quality factor* of a resonator is the mode frequency divided by the line width; $Q = \dfrac{\omega_m}{(\delta\omega)_{1/2}}$.

- The number of oscillations of the electromagnetic field before the energy drops to $1/e$ of its original energy is $N_{1/e} = \dfrac{Q}{2\pi}$.

- The *free spectral range* (FSR) is the difference in angular frequency between adjacent modes; $\Delta\omega = \omega_{m+1} - \omega_m = \omega_m/m$.

- The *finesse* is the FSR divided by the line width,

$$F = \Delta\omega/(\delta\omega)_{1/2} = Q(\Delta\omega/\omega_m) . \qquad (6.46)$$

Since $\Delta\omega/\omega = 1/m$, the *finesse* is just $F = Q/m$. Letting $Q = 2\pi N_{1/e}$, the *finesse* is seen to be 2π times the number of round trips $N_{r.t.}$ before the energy decays to $1/e$ of its original value; $F = 2\pi N_{r.t.}$.

6.4 – Whispering Gallery Mode (WGM) Resonator and Sensor: light confinement by total internal reflection

Having a resonator with a very large quality factor Q enables a great many technological innovations such as sensors that can measure and/or detect extremely small forces, temperature changes, refractive index changes, viruses, single molecules, etc. The latter is something I became interested in after observing in 1994 that a certain type of resonator maintained a high Q (~10^6) after being immersed in aqueous solution (Serpenguzel, Arnold, Griffel, 1995). At the time, I suggested it as a candidate for an ultrasensitive biosensor and estimated that it would be sensitive to a layer of adsorbed atoms having a thickness of 10^{-11} m; one-tenth the diameter of a hydrogen atom. My estimate was based on a calculation of the frequency shift of the resonant modes caused by the binding of the layer. The implication was that a small fraction of a monolayer of protein could be detected, since protein molecules are considerably larger than atoms. The resonator at the time was a microsphere that contained orbiting light confined by total internal reflection (Fig. 6.13a). It was driven into resonance by coupling it evanescently to light traveling through an adjacent optical fiber. We said earlier that total internal reflection is without loss at a flat surface, and although that is not the case for a curved surface, the losses can be very small, which allows the Q to be large. The experiment performed in my laboratory [**MicroParticle PhotoPhysics Lab (MP³L)**] at that time is displayed on the cover of this book. As you can see the resonances correspond to narrow spikes in detected power, at specific wavelengths.

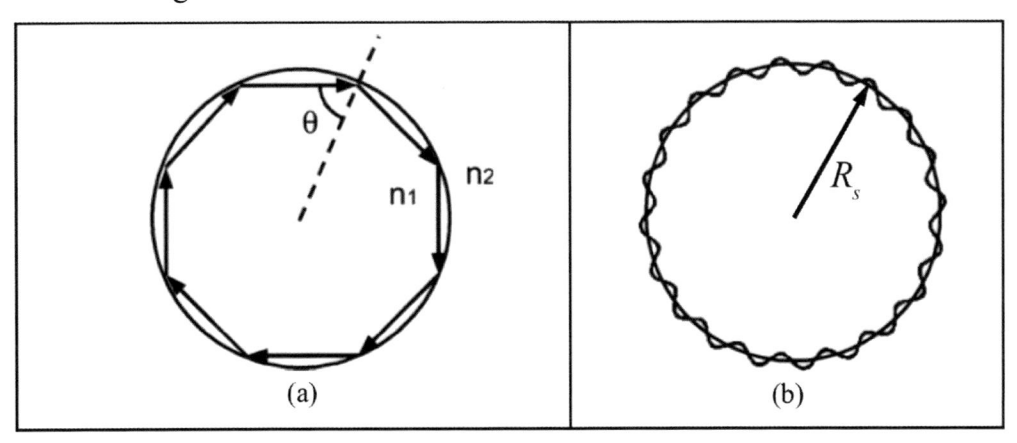

Fig. 6.13 – The two faces of a Whispering Gallery Mode. (a) Light rays confined by total internal reflection in a polygon orbit, and (b) Waves that wrap around the equator and return in phase.

The circulating light is known as a Whispering Gallery Mode (WGM) because it has an acoustic analog. There are circular architectural structures around the world where you can whisper near the wall and be heard by a friend diametrically opposed, and tens of meters away. Perhaps the most famous of these is in London, within St. Paul's cathedral. Here the radius of the gallery is nearly 20 meters, still a receiver diametrically opposed will hear your whisper.

The optical WGM is much smaller; it can be only a few microns in size. This is principally because the speed of sound is so much smaller than the speed of light. The optical WGM is described in Fig. 6.13 by its two faces. The figure on the left clearly demonstrates why the light is confined, while the figure on the

159

right explains resonance. Venturing into the details of these orbits is beyond the level for which this book was intended, however the simple picture on the right provides the ability to write down some intuitive equations. Let's start by calling the wavelength of the propagating light λ_m. Then it appears that an integer m of these wavelengths add up to the circumference of an orbit,

$$2\pi R_s = m\lambda_m . \qquad (6.47)$$

The integer m is the mode designator in our simple model. If one were to track the mth mode as a layer having similar dielectric properties is added to the radius then the wavelength of the resonator will shift by

$$\Delta\lambda_m = \frac{2\pi}{m}\Delta R_s = \lambda_m \frac{\Delta R_s}{R_s} , \qquad (6.48)$$

where we have used Eq. (6.47) for the right-hand most part of Eq. (6.48). One can turn Eq. (6.48) around to ask how small a layer can be detected from a discernible wavelength shift. An easily discernible shift would have to be comparable to the line width. If we replace the $\Delta\lambda_m$ by the line width $\delta\lambda_m$ in Eq. (6.48), then the layer thickness needed to produce a shift of one line width is

$$\left(\Delta R_s\right)_\delta = R_s \frac{\delta\lambda_m}{\lambda_m} . \qquad (6.49)$$

As long as the fractional wavelength shift is small, $\delta\lambda_m/\lambda_m$ is simply the inverse of the quality factor, $1/Q$, and the layer thickness required to shift a resonance by a line width is

$$\left(\Delta R_s\right)_\delta = \frac{R_s}{Q} . \qquad (6.50)$$

The particles we looked at in 1994 had Q's of about 10^6 and radii down to 10 microns. This translates into a layer thickness of only 10^{-11} m; one-tenth the diameter of a hydrogen atom! A typical protein molecule is $300\times$ larger than this. No wonder then that an industry is growing around the Whispering Gallery Mode biosensor.

Scan this QR code to learn more about the WGM biosensor.

6.5 – Chapter Six Exercises

Exercise 6.1 – A scattered intensity spectrum such as the one recorded in Fig. 6.2 for a silver particle in air (e.g. red curve) can be used to estimate how long the energy built up in the particle takes to decay after the driving field is suddenly turned off. **Estimate the time to decay to 1/e of the energy at cut-off.**
Hint: The scattered intensity is proportional to the absolute polarizability squared; $|\alpha_{LSP}|^2$, which means that the full width at half maximum is just γ, the plasmon relaxation rate.

Exercise 6.2 – Show that Eq. (6.8) and Eq. (6.9) are equivalent.

Exercise 6.3 – Plot Eq. (6.17) for $R = 0.9$ to see if you get a pronounced resonance as in Fig. 6.6. Hint: you will have to change the phase constant δ by a bit more than π. Try 99.5π to 100.5π.

Exercise 6.4 – Put the resonance you find in Exercise 6.2 into a Lorentzian form. Is it a perfect Lorentzian or just approximate?

Exercise 6.5 – Use Gauss's law to show that the field inside a uniform positively charged sphere is proportional to the displacement from the center y_e and to the density of positive charge ρ_+, i.e.

$$E = \rho_+ y_e / (3\varepsilon_0).$$

Exercise 6.6 – Explain why it is possible for all incident power on an etalon to be transmitted, even though each of its two sides reflects light.

Exercise 6.7 – Explain why the line-width of the resonance in Fig. 6.6 is affected so profoundly by the reflectance R at either side of the resonator.

Exercise 6.8 – Show that Equations (6.17) and (6.33) are equivalent.

Exercise 6.9 – Plot Eq. (6.33) for an etalon as a function of $\varphi = nk_0 d$ for refractive indices of 1.5, 3, and 5, from $\varphi = 200$ to $\varphi = 210$.

Exercise 6.10 – Calculate the power spectrum of the damped oscillator $(a > 0)$:

$$f(t) = \begin{cases} Ae^{-(a-i\omega_0)t}, & t \geq 0 \\ 0, & t < 0 \end{cases}.$$

Exercise 6.11 – Show that the power spectrum of a Gaussian pulse

$$f(t) = A\, e^{-at^2 + i\omega_0 t}$$

is also a Gaussian function centered at ω_0.

Exercise 6.12 – Imagine a cavity with both sides having reflectivity R_2, i.e. both sides allow photons to escape. Find the lifetime τ for the system and use it to estimate the line width $(\delta\omega)_{1/2}$.

Exercise 6.13 – Estimate the mode numbers m for the narrowest resonances in the spectrum on the book cover. In particular, what is m for the resonance at 606 *nm*? Hint: Construct an equation for m in terms of this central wavelength λ_m and the free spectral range $(\lambda_m - \lambda_{m+1})$, and take measurements from the spectrum on the book cover in order to obtain a numerical value for m.

Exercise 6.14 – The free space wavelength used to excite each Whispering Gallery Mode (WGM) for the polystyrene micro-particle on the cover is not the wavelength within each WGM as depicted in Fig. 6.13(b). One might think that the internal WGM wavelength is just the free space wavelength $\lambda_{0,m}$ divided by the refractive index of polystyrene. However this is not the case because each WGM has an evanescent field that dips into the water. Consequently we must define an effective refractive index n_{eff} and modify Eq. (6.47) to read

$$2\pi R_s = m \frac{\lambda_{0,m}}{n_{eff}}. \tag{6.51}$$

Given that the microsphere on the cover has a radius $R_s = 11.8\,\mu m$, determine the effective refractive index of the WGMs.

Chapter Seven – LASER

7.1 – What would happen if a Fabry-Perot etalon had positive gain?

So far we have not considered losses of photons within the cavity, for example by absorption, or the possibility that the opposite could occur – the generation of new coherent photons. The latter occurs through stimulated emission from excited atoms or molecules. Fortunately we are in a good position to incorporate such "gain" because of our development of Etalon theory. The device that arises from such theory is the LASER (**L**ight **A**mplification by **S**timulated **E**mission of **R**adiation).

To begin let's recall Eq. (6.16) for the transmitted field from the F-P etalon

$$\frac{E_t}{E_0} = \frac{(1-r^2)e^{i\delta}}{1-r^2 e^{i2\delta}}. \tag{7.1}$$

Suppose there are losses in each single pass through the etalon. A way to account for these losses is to include a complex phase, $\delta = (k + i\alpha)d$, where $k = n_r k_0$. In so doing there is an attenuation of the field as the light moves from one reflecting surface to the next by a factor of $e^{-\alpha d}$. However if α is negative, then there is gain; negative attenuation. It is convenient in this situation to express α as $-|\alpha_g|$. The effect that this has on the transmitted light at a peak in transmission (i.e. maximum) is profound,

$$T = \frac{I_t}{I_0} = \left|\frac{E_t}{E_0}\right|^2 = \frac{(1-R)^2 e^{2|\alpha_g|d}}{\left(1-Re^{2|\alpha_g|d}\right)^2}, \tag{7.2}$$

where the substitution of $R = r^2$, is valid when r is real. Instead of the transmitted intensity at resonance (i.e. $kd = m\pi$) being capped at the incident intensity, as it is when $|\alpha_g| = 0$, the transmission grows well beyond 1 with gain. Fig. 7.1 shows a plot of Eq. (7.2) for 0.3 mm thick cavity having reflectivity $R = 0.30$. Although with $|\alpha_g| = 0$, $T = 1$, as the gain increases the light exiting the structure is remarkably increased. An actual laser is more complicated, but

163

before dealing with such details, we will spend sometime discussing the mechanism for this gain.

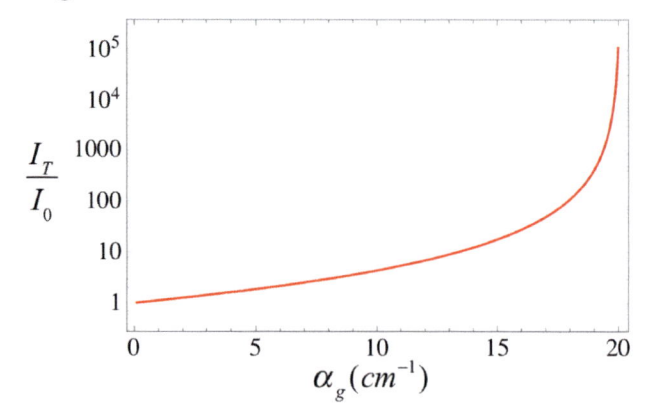

Figure 7.1 – Effect of gain on etalon transmission on resonance.

7.2 – Stimulated Emission

The gain associated with lasing requires quantum mechanics to be fully described. It is the result of an interaction between photon and atom or molecule having quantized energy levels. The simplest way to look at this is to imagine two levels within an atom,[18] an upper state with energy ε_2 and the lower state with energy ε_1. If we suppose the atom is initially excited as in Fig. 7.2, then a photon incident on it with energy $hf = \varepsilon_2 - \varepsilon_1$ causes a strange and rather wonderful effect. The atom emits another photon that is a clone of the incident photon. By clone I mean that it has the same phase, frequency, polarization and direction. If this happens trillions of times, an intense beam of coherent light is produced at a single frequency.

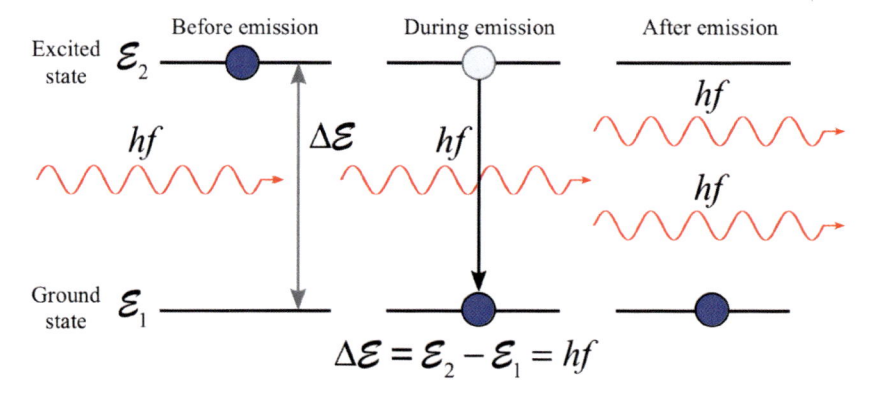

Figure 7.2 – Generation of photons via stimulated emission.

---

[18] To make the description simple we have taken the ground and excited states to be one state each. Such states are called non-degenerate.

Until the laser came along a single frequency, coherent visible light source was unavailable. Instead visible light was overwhelmingly produced through *spontaneous emission*, as is the case with incandescent light bulbs and fireflies. It is light for which stimulation by a photon is not necessary, and the spontaneously emitted photons have random phase, and polarization. Lasing which arises from stimulated emission comes at a price. Electrons must be pre-excited electrically or optically. The first stimulated emission event requires a seed photon that typically starts from a spontaneously emitted photon.

Einstein first anticipated stimulated emission in the year 1916. He found that he could not explain the radiation spectrum from an almost perfect absorber without having both spontaneous and stimulated emission. His theory laid the groundwork for understanding the laser, although the laser was invented more than 40 years later. His argument was simple and brilliant. To get started I will describe Planck's radiation law.

The radiation intensity spectrum per unit frequency $I(f)$ from an almost perfect absorber held at temperature in Kelvin, T_K, accurately matches Planck's radiation law (for which Planck received a Nobel Prize),

$$I(f) = \frac{2\pi h f^3}{c^2} \frac{1}{e^{hf/kT_K} - 1}, \qquad (7.3)$$

where f is the light frequency, and k is Boltzmann's constant. Planck fit an equation of this form to experimental data, and then through a hypothesis of quantized energy levels he arrived at a mechanism for explaining Eq. (7.3). Planck's theory made no distinction between spontaneous and stimulated emission. Einstein took a different approach. In his paper, "On the Quantum Theory of Radiation", he introduced stimulated emission. Next we will say more about the perfect absorber that led to Eq.7.3, before we delving into Einstein's interpretation.

To be certain that the problem was well defined the researchers chose a surface that was completely absorbing. Fig. 7.3 shows their standard called a blackbody.

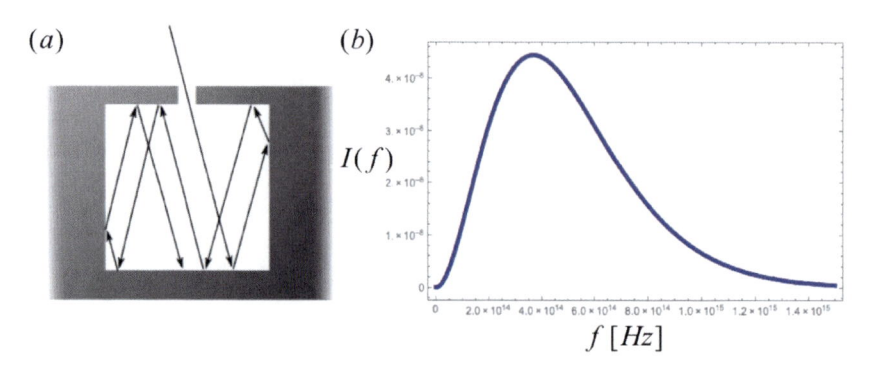

Figure 7.3 – *(a) Idealized blackbody. (b) Radiated intensity Eq. (7.3) as a function of frequency for a blackbody with $T = 6000\,K$.*

By digging a cavity below a surface that breaches the surface with a small exposed hole (the surface element), the probability for absorption by the surface of the hole is 1. What is emitted from the hole is the result of incandescence due to heating the cavity and the other material it is embedded in. Although the experimenters were interested in Eq. (7.3), which described the intensity emitted, Einstein's interest was in the intensity of light generated in the interior, which has a spectral energy density $\rho(f)$,[19]

$$\rho(f) = \frac{8\pi h f^3}{c^3} \frac{1}{e^{hf/kT_K} - 1} . \qquad (7.4)$$

First Einstein recognized that a statistical number of atoms in the walls of the cavity contributed to the problem. At temperature T_K there is a probability that the upper level in Fig. 7.2 can be occupied. Based on Boltzmann's equilibrium equation, that ratio of the number of electrons in the upper level to the number in the lower level is

$$\frac{N_2}{N_1} = e^{-\frac{(\varepsilon_2 - \varepsilon_1)}{kT_K}} . \qquad (7.5)$$

To construct the emitted intensity Einstein could have assumed that only two things occur to change the population in level 2: absorption of photons to raise electrons from $1 \to 2$ with rate $B_{1\to 2} \cdot \rho(f) \cdot N_1$, and spontaneous emission with rate $-A \cdot N_2$. This leads to an overall rate of change in the number of photons in level 2

---

[19] $\rho(f)$ has units of energy per unit volume per unit frequency.

$$\frac{dN_2}{dt} = B_{1\to2}\rho(f)N_1 - AN_2 \ . \tag{7.6}$$

In equilibrium $\frac{dN_2}{dt} = 0$, so $\rho(f) = \frac{A}{B_{1\to2}} \frac{N_2}{N_1}$. Combine this with Eq. (7.5) and you will have

$$\rho(f) = \frac{A}{B_{1\to2}} \frac{1}{e^{hf/kT_K}} \ . \tag{7.7}$$

By comparing Eq. (7.7) to Eq. (7.4) he would have realized that spontaneous emission alone would not work. Then he had his brilliant idea. If a photon could induce absorption from state 1 to 2, perhaps it might also induce emission from state 2 to 1. Thus, he revised Eq. (7.6)

$$\frac{dN_2}{dt} = B_{1\to2}\rho(f)N_1 - B_{2\to1}\rho(f)N_2 - AN_2, \tag{7.8}$$

which gives an equilibrium intensity

$$\rho(f) = \frac{A}{B_{1\to2}e^{hf/kT_K} - B_{2\to1}} \ . \tag{7.9}$$

In examining Eq. (7.9) he could see a clear path to Planck's equation. First he postulated symmetry between induced absorption and induced emission by setting $B_{1\to2} = B_{2\to1} = B$. Next by returning to Eq. (7.4) he identified the ratio $A/B = 8\pi h f^3/c^3$.[20] His guess of this new induced (aka stimulated) emission completed the story and inspired the technologies of coherent light amplification and the laser, which combines a light amplifier with feedback. The symbols A and B are revered as the Einstein A and B coefficients. Once having A and B we can go on to the laser.

At first you might think that all we need is a volume of atoms and irradiate them as intensely as possible. That would work if the atoms could all be driven into their upper state, and although you might think that a blast of absorbing energy is all that would be needed, with a two state atom it won't work so long as

---

[20] Einstein's analysis was done in terms of the energy density within a blackbody cavity, for which he obtained $A/B = 8\pi h f^3/c^3$. Note that the units are energy per unit volume per unit frequency.

equilibrium is to hold (Eq. (7.5)). That is because of the symmetry of the B coefficients. Beyond the time at which half of the electrons are in the upper state and half are in the lower state, no further imbalance can occur that favors the upper state.

Perhaps a better way to say the same thing is to look at the ratio of the rate of emission to the rate of absorption,

$$\frac{|\text{Rate of Emission}|}{|\text{Rate of Absorption}|} = \frac{B\rho(f)N_2 + AN_2}{B\rho(f)N_1} = \frac{N_2}{N_1}\left(1 + \frac{A}{B\rho(f)}\right). \quad (7.10)$$

If we now increase $\rho(f)$ in order to make $B\rho(f) \gg A$, ultimately what limits our ability to make the rate of stimulated emission greater than the rate of absorption in an equilibrium situation is that N_2/N_1 must be less than 1. To create a laser this ratio must be reversed; we must operate the laser out of equilibrium, where we have the possibility of population inversion, $N_2/N_1 > 1$.[21]

7.3 – Original Laser and Other Types

Theodore Maiman demonstrated the first laser in 1960. He used a three state system (Fig. 7.4).

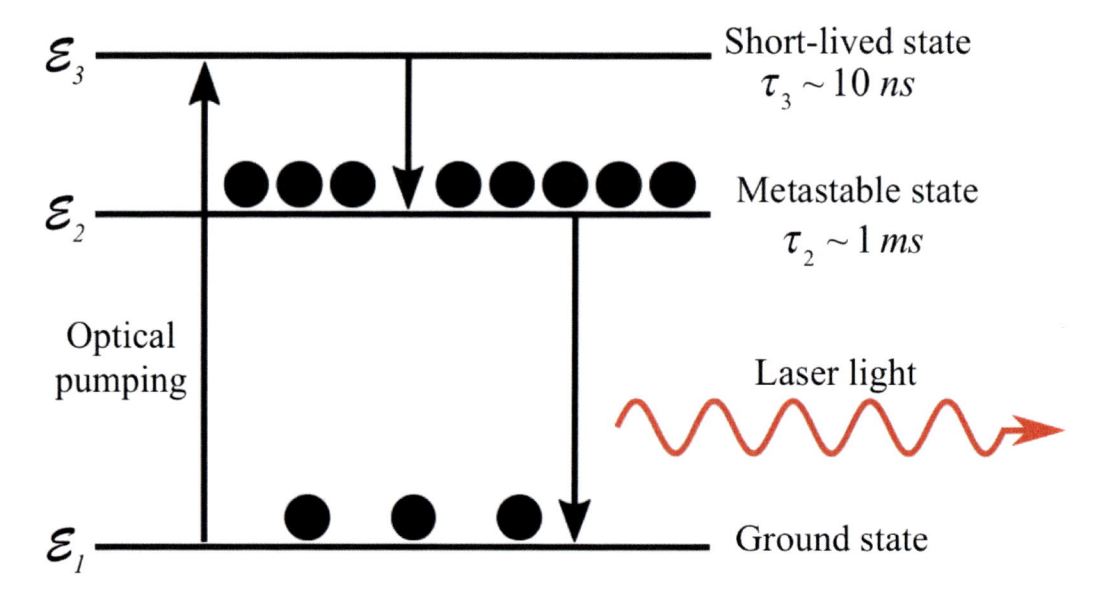

Figure 7.4 – Three state system.

---

[21] Although we will see that population inversion can be achieved with a three level system, rapid spontaneous emission will always be a problem. The Einstein A/B ratio is telling in this regard since it goes as frequency cubed; x-ray lasers are very hard to design.

The idea is that the electron is driven to a state from which there is a rapid decay into a lower, longer-lived, metastable $\mathcal{E}_2$ state. By metastable we mean a state for which spontaneous emission to the ground state is inhibited. All that means usually is a state for which quantum mechanics restricts the interaction to ground state. So unlike the state $\mathcal{E}_3$ in Fig. 7.4 where the electron is originally pumped, and which has a lifetime on the order of 10 nanoseconds the state $\mathcal{E}_2$ has a lifetime in milliseconds; for milliseconds there can be a population inversion. The first spontaneously emitted photon will find many electrons in the metastable stable state and begin an avalanche of stimulated emissions.

Maiman's material system was Ruby. Ruby is a gemstone but is also produced synthetically. It consists of a chromium impurity inside an aluminum oxide crystal. Crystalline aluminum oxide is known as sapphire. Figure 7.5 shows Maiman's laser. The pump from state $\mathcal{E}_1$ to $\mathcal{E}_3$ is provided by a flash-lamp. To capture additional light from the flash-lamp it is housed in a mirrored cylinder. On a historic day in 1960 after a short pulse of light from the flash-lamp, a red dot was seen in the distant emission indicating that lasing had occurred.

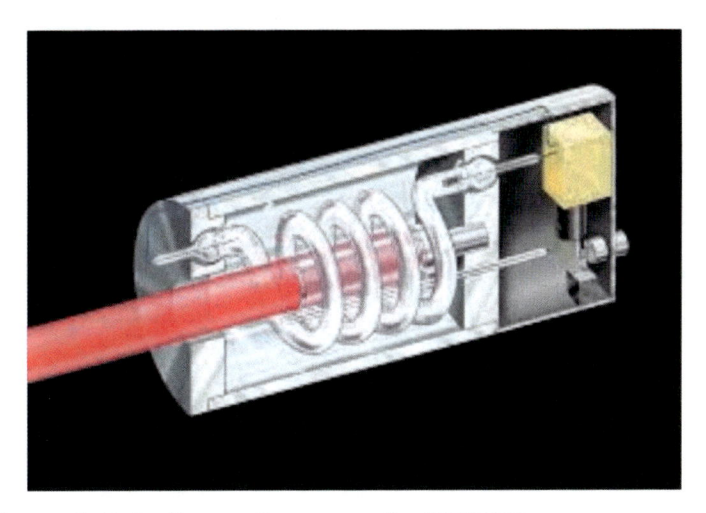

Figure 7.5 - Maiman's Ruby Laser. (Image credit: USDOE)

There are many types of lasers, but the most common, shown in Fig. 7.6, is the Laser Diode. As its name implies it is diode made of a semiconducting sandwich, with excess electron acceptors (P-type material) on one side, and excess electron donors (N-type material) on the other side. The electron acceptors lead to mobile positive charges (holes), while the electron donors lead to free electrons. Between the two is the intrinsic material (i.e. undoped). Within this intrinsic thin layer free electrons and holes combine, and spontaneously emit. At this point the device is a light emitting diode (LED). The continuous generation of new free electrons and holes from the external current source provides a gain process, and reflection from the right and left sides creates a Fabry-Perot cavity. These elements combine to form a very compact laser. A spectrum of the emitted light is shown in Fig. 7.7.

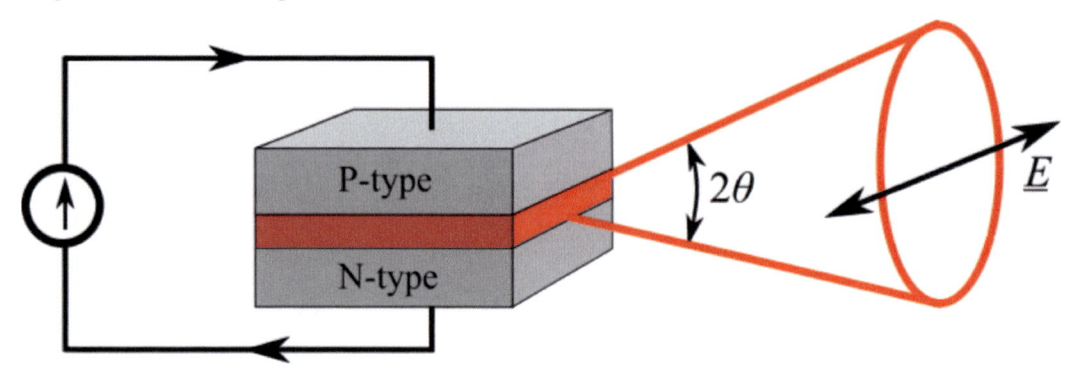

Figure 7.6 – Laser Diode construction and emission.

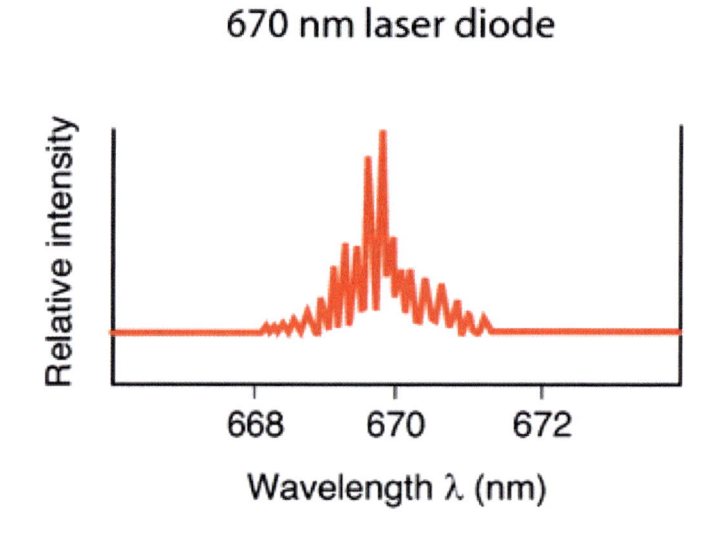

Figure 7.7 – Spectrum from a diode laser. (Image credit: Newport Corporation)

This spectrum has many separate peaks (~7.0 modes per nm) because the semiconductor has a very broad gain bandwidth that overlaps many longitudinal modes. You can imagine these modes as if they were standing wave modes on a string (Fig. 7.8).

170

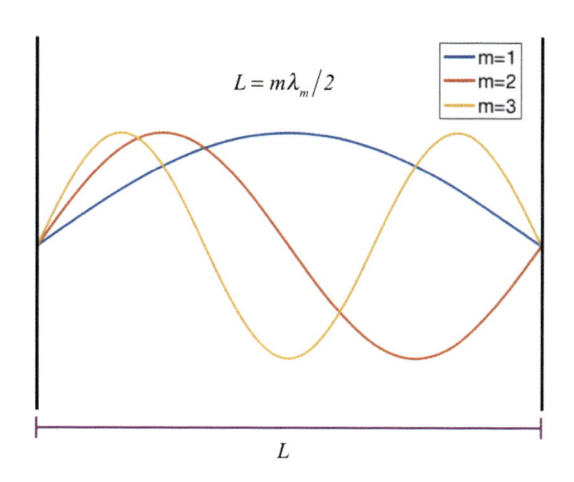

Figure 7.8 – String modes.

The string displacement is an analog of the transverse electric field in the laser, and λ_m is the internal wavelength of the light. When the internal wavelength is twice the cavity length L the lowest frequency mode is excited; $m=1$. The $m=1$ mode has half the wavelength between the front and back of the resonator. The modes in Fig. 7.7 are so closely spaced in wavelength that their mode numbers must be very large. The closure condition for having resonant modes is stated simply as - "a resonance has an integer number of half wavelengths that fit into the cavity length L";

$$L = m\lambda_m/2. \tag{7.11}$$

This last equation will be known as the "closure condition" for resonance. Clearly the modes excited in Fig. 7.7 are of much higher order since they are so closely spaced. What is really neat about this spectrum is that it can be used to obtain the cavity length L.

Since L is fixed for a given laser if we think about going from a given mode m to the next mode $m+1$, we can take differentials on either side of Eq. (7.11),

$$2\Delta L = 0 = \Delta m \lambda_m + m(\lambda_{m+1} - \lambda_m). \tag{7.12}$$

With $\Delta m = 1$, we see that m is simply

$$m = \frac{\lambda_m}{\lambda_m - \lambda_{m+1}}. \tag{7.13}$$

It should be pointed out that each λ_m is less than the corresponding free-space wavelength λ_{m0}; $\lambda_m = \lambda_{m0}/n$, where n is the refractive index of the cavity. However, Eq. (7.13) still holds true with each of the wavelengths replaced by free space wavelengths. With λ_{m0} taken to be 670 nm, and the spacing between modes taken as (1/7) nm, m is estimated to be 4890. To get the cavity size L we return to Eq. (7.11) with λ_m replaced by $\lambda_m = \lambda_{m0}/n$. The material from which the "cheap" laser is fabricated is Gallium Indium Phosphide (GaInP) with a refractive index of ~3 for wavelengths near 670 nm. From Eq. (7.11), L is found to be 546,050 nm or approximately 0.55 mm. That's really small!

7.4 – Chapter Seven Exercises

Exercise 7.1 – Atoms emit energy in discrete chunks or "quanta" that are called photons. In chapter one we identified the energy of a photon as

$$\varepsilon = \hbar\omega.$$

What is the energy of a photon with $\lambda = 500 \ nm$?

Exercise 7.2 – Blue light with wavelength $\lambda_{0,b} = 458 \ nm$ is used to excite fluorescent molecules. If the fluorescent molecules emit green light with an average wavelength $\lambda_{0,g} = 532 \ nm$ what amount of energy was lost to molecular vibrations for each of the blue photons that excited the sample? Recall: $\hbar = h/2\pi = 1.05 \times 10^{-34} \ J \cdot s$.

Exercise 7.3 – Electrons in a three-state system are pumped from the ε_1 state to the ε_3 state through the absorption of green light (545 nm). Electrons then reach the ε_2 through a non-radiative process involving crystal vibrations. This leads to the population inversion shown in the figure. Next a spontaneous emission from ε_2 to ε_1 stimulates an avalanche of coherent photon emissions, as electrons are returned to ε_1. Each photon in this LASER beam has energy of $2.86 \times 10^{-19} \ J$ or $1.79 \ eV$.

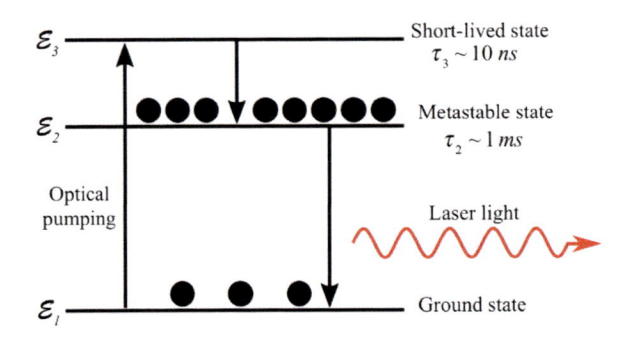

(a) What is the energy difference between states $\mathcal{E}_1$ and $\mathcal{E}_3$ in Joules?

(b) What is the wavelength of the LASER emission in nanometers?

Planck's constant $h = 6.6 \times 10^{-34} J \cdot s$.

Exercise 7.4 – Use Eq. (7.3) to show that the intensity per unit wavelength of blackbody radiation is

$$I(\lambda) = \frac{2\pi h c^2}{\lambda^5} \frac{1}{e^{hc/(\lambda kT)} - 1}. \tag{7.14}$$

Hint: Because frequency and wavelength are linked through $c = \lambda f$ the intensity per unit frequency over a span of frequency df is the same as the intensity per unit wavelength over the corresponding span in wavelength $d\lambda$;

$$I(f)df = -I(\lambda)d\lambda. \tag{7.15}$$

Exercise 7.5 – *Wien's Law* – Use Eq. (7.14) to find an equation for the maximum wavelength radiated by a blackbody at a given temperature. You should arrive at an expression of the form

$$\lambda_{max} = \frac{\eta}{T},$$

where η is a constant with units of $nm \cdot K$ (K is degrees Kelvin). Hint: take the derivative of Eq. (7.14) with respect to λ and set it equal to 0.

You will need some constants:

Planck's constant: $h = 6.63 \times 10^{-34} J \cdot s$ (Joules-sec), $\hbar = h/2\pi = 1.05 \times 10^{-34} J \cdot s$.

Boltzmann's constant: $k = 1.38 \times 10^{-23} J/K$ (Joules/Kelvin).

Exercise 7.6 – Use the result obtained in Ex. 7.5 to find the maximum wavelength emitted form the following:

  (a) The Sun – $T = 5,778 K$.

  (b) Your skin temperature – $T = 307 K$.

Exercise 7.7 – The cosmic microwave background (CMB) radiation is measured to have a peak wavelength of $\lambda = 1.063 mm$. Using this wavelength determine the average temperature of the visible universe.

Experiment #7 Structure of diode laser from far-field diffraction

Unscrew and remove the lens at the front of your laser provided in the Pocket Optics kit. Carefully place the lens and enclosed spring aside so you can replace them later. Turn on your laser and observe the laser intensity pattern in the absence of the lens. The light is no longer collimated but spreads out broadly. The spread of the light is caused by diffraction, a subject you are now familiar with from Chapter 2. By studying the diffracted intensity pattern we can learn something about the aperture that the laser light is emitted from. Suppose that the laser light exits the diode through a rectangular aperture that is created by an intrinsic layer of semiconductor with width t, as shown in Fig. 7.6. Assume that the length of the rectangle is sufficiently large in comparison to its width so that the aperture can be treated as a single slit. Use your photometer photodiode to measure the light intensity at the center of the scattering pattern where it is at a maximum, and then move your photometer board away from center to find where the intensity falls to half its maximum value. Use the measurement and your knowledge of diffraction to estimate the slit width t of the emitting aperture. The scattering pattern is actually Gaussian in shape rather than sinc squared, but this approximation gives reasonable enough results for our purposes.

A possible set-up using the Pocket-Optics kit is shown below. I have used the G slot as a guide for the photometer board, and seated the laser as far back as possible within its cradle on the Baseboard. The distance from the laser source to the photometer card is about 38 mm. By taping the metal ruler to the base plate along the G slot you can use the ruler to measure how far the photometer board is moved to go from the maximum to half intensity.

Experiment#7 Setup

Scan for a video demo of this experiment

175

Chapter Eight – Spectroscopy

8.1 – Grating Spectrometer

In the last Chapter we discussed the theory that led to the development of the laser. You may remember a figure from that chapter that gave the spectrum of a diode laser. The figure has been reproduced in Fig. 8.1, and in this section we will examine how measurements such as this are made.

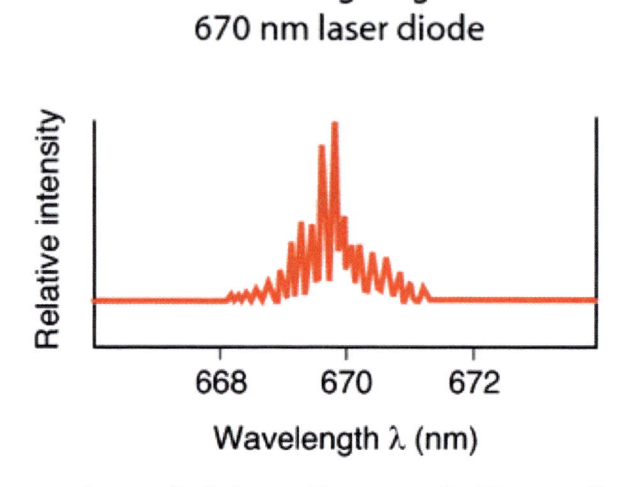

Figure 8.1 – Spectrum from a diode laser. (Image credit: Newport Corporation)

The spectrum in Fig. 8.1 was likely taken with a "grating-spectrometer" using 1st order diffraction. Actually you may be able to make an adequate spectrometer yourself using a simple holographic diffraction grating. Fig. 8.2 shows how this might be possible. The key question: can the individual peaks in Fig. 8.1 be resolved?

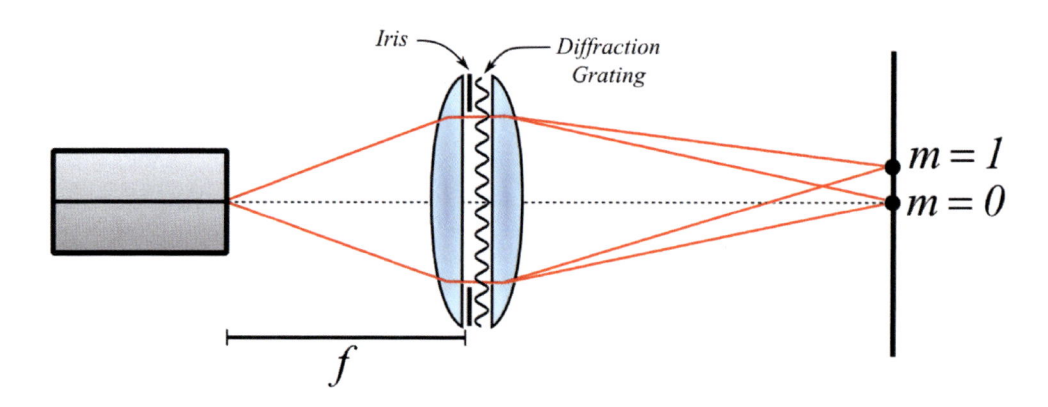

Figure 8.2 – Transmission grating spectrometer.

For the spectrum in Fig. 8.1 this requires a resolution better than 0.14 nm since there are 7 modes per nm; take the time to count them. Such a resolution is quite small. Therefore before describing the details of the grating spectrometer, it is important to discuss the physics that governs resolution. It turns out that the number of grating "slits" that the light illuminates, N, is the key.

As you will recall, the holographic diffraction grating described in Chapter 2 can only produce three diffracted beams. They are two first orders, and a zeroth order. The zeroth order contains no information about the wavelengths/frequencies that the laser emits. This information is contained in the first order diffraction (e.g. $m=1$). So being able to separate two closely spaced beams in wavelength/frequency involves seeing two distinct peaks in the first order diffraction. If one of the first order beams is described by the delta function $\delta(k_{1y} - 2\pi/P)$ as we did in Eq. (2.53), it may give the correct angle of the first order diffraction peak $sin(\theta_1) = k_{1y}/k = \lambda/P$, but the delta function's zero width is misleading; the beam actually has a width δk_{1y}, and this width limits the resolution. The problem with the delta function is that it assumes an infinite number of "slits" are illuminated, which is unrealistic. In what follows we will return to the case of a finite number of slits N, but preserve the same pitch P with the goal of determining δk_{1y}.

To get a general result we consider the case in which the diffraction grating is between the cylindrical lenses in Fig. 8.2, and contains an iris so that only N slits are exposed to the light. If we represent the grating as $A_g(y) = 1/2 - (1/2)\cos[2\pi y/P]$, and the iris exposes an even number N of "slits", then the one-dimensional Fourier transform describing the field for a deviation k_y is

$$c(k_y) = \int_{-NP/2}^{NP/2} \left[1/2 - (1/2)\cos(2\pi y/P)\right] e^{-ik_y y} dy, \qquad (8.1)$$

which evaluates to be

$$c(k_y) = \frac{NP}{2}\text{sinc}\left(\frac{k_y NP}{2}\right) - \frac{NP}{4}\left[\text{sinc}\left(N\pi - k_y NP/2\right) + \text{sinc}\left(N\pi + k_y NP/2\right)\right], \qquad (8.2)$$

where the sinc function is defined by $\text{sinc}(\xi) = \sin(\xi)/\xi$. The three terms consist of the 0th order (first term), and the two 1st orders. The first order that produces the beam with a positive k_y is the 2nd term,

$$c_1(k_y) = -\frac{NP}{4}\left[\text{sinc}\left(N\pi - k_y NP/2\right)\right]. \tag{8.3}$$

This field amplitude peaks when the argument of the *sinc* function is zero. That occurs when $k_{1y} = 2\pi/P$, which for perpendicular incidence on the plane of the grating is the same as the spatial frequency of the grating k_g. By setting $\Delta k_y = k_y - k_g$ in Eq. (8.3) we have a coordinate that is centered on this peak. Rewriting the argument in Eq. (8.3), and computing the intensity of the first order peak gives

$$\left|c_1(k_y)\right|^2 = \left(\frac{NP}{4}\right)^2 \text{sinc}^2\left(\frac{\Delta k_y NP}{2}\right). \tag{8.4}$$

We can evaluate the width of this the primary peak in this function by finding the smallest value of Δk_y that makes $\left|c_1(k_y)\right|^2 = 0$. This occurs when the argument of the *sinc* function is π, for which the diffraction broadened line width component from Eq. (8.4) is $\delta k_{1y} = 2\pi/(NP)$. Although the beam's y-component k_{1y} is unaffected by the number of slits, its width is proportional to $1/N$. Our original question was concerned with the smallest wavelength difference that can be resolved between two lines having different wavelengths. We are now in a good position to answer that question.

Our two main results from the previous discussion are that the y-component of the wave-vector of the scattered wave k_{1y} and the broadening due to diffraction δk_{1y} are

$$k_{1y} = \frac{2\pi}{P}, \text{ and} \tag{8.5}$$

$$\delta k_{1y} = \frac{2\pi}{NP}. \tag{8.6}$$

We will use these equations to determine the wavelength resolution. From an experimental point of view what counts is what is recorded on a detector. For light having wavelength λ, the recording medium records an angular record as shown in Fig. 8.3. The 1st order diffracted "spot" has a position θ_1 and a width $\delta\theta_1$. If light at a longer wavelength $\lambda + \Delta\lambda$ is to be discriminated from the light at wavelength λ, it must have a diffracted angle $\theta_1 + \Delta\theta_1$ no smaller than $\theta_1 + \delta\theta_1$, which makes the smallest deviation in angle that can be discriminated $(\Delta\theta_1)_{min} = \delta\theta_1$. Eq. (8.5) can be rewritten as $P\sin(\theta_1) = \lambda$, so that by differentiating we find that $P\cos(\theta_1)(\Delta\theta_1)_{min} = \Delta\lambda$. From Eq. (8.6) we find that $P\cos(\theta_1)\delta\theta_1 = \lambda/N$. With the latter equation divided into the former, and using $(\Delta\theta_1)_{min} = \delta\theta_1$, we obtain the resolution limit in wavelength

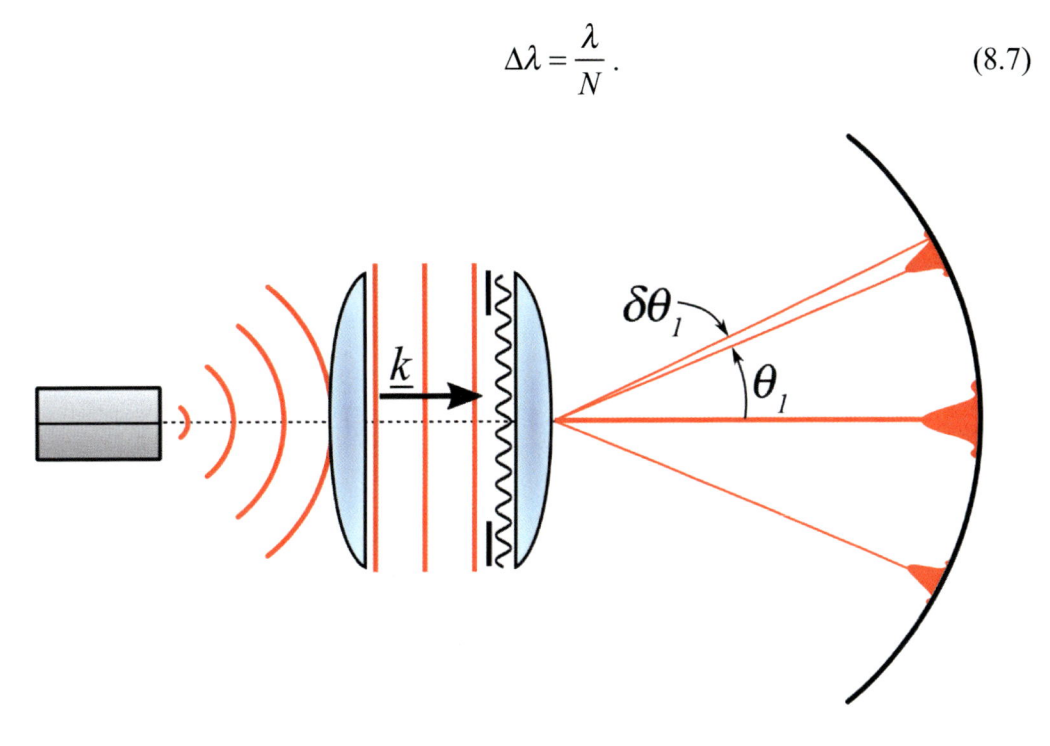

$$\Delta\lambda = \frac{\lambda}{N}. \qquad (8.7)$$

Fig. 8.3 – Beams resulting from diffraction of monochromatic light by a grating. The detection film is at one focal length, f, from the 2nd lens. The figure is illustrative and compressed horizontally. Note that the width of the first order maximum is illustrated as being smaller than the central maximum. Is this as expected?; see Ex.8.1.

As an example specific to your own holographic diffraction grating, if you were somehow able to illuminate all of the 25,400 "slits" with a laser having a nominal wavelength of 670 nm, then the resolution could theoretically be as small as $670 \ nm/25,400 = 0.027 \ nm$. That is smaller than the estimated separation between the laser modes by a factor of $0.14 nm/0.027 nm \sim 5$. So, using a round number of 10,000 "slits" would probably be O.K. It is important to note that the typical 2 mm beam width from your laser pointer "as is" would not allow the individual laser modes to be resolved. The typical laser pointer's collimated beam width is ~2mm, allowing only ~2000 graing slits to be illuminated.

8.2 – Fourier Transform Infrared Spectroscopy (FTIR) and the Michelson Interferometer

Michelson interferometer is named for Albert Michelson who used the device to test for the medium that many scientists before 1900 thought was responsible for light's motion. They called the medium the aether. Although Michelson found no evidence for the aether, consistent with Einstein's postulate, the interferometer he invented became a major tool for molecular spectroscopy.

Fig. 8.4 shows the basic interferometer. It consists principally of two mirrors, M_u on top and M_s on the side, a 50:50 beam splitter at the center, and a detector d.

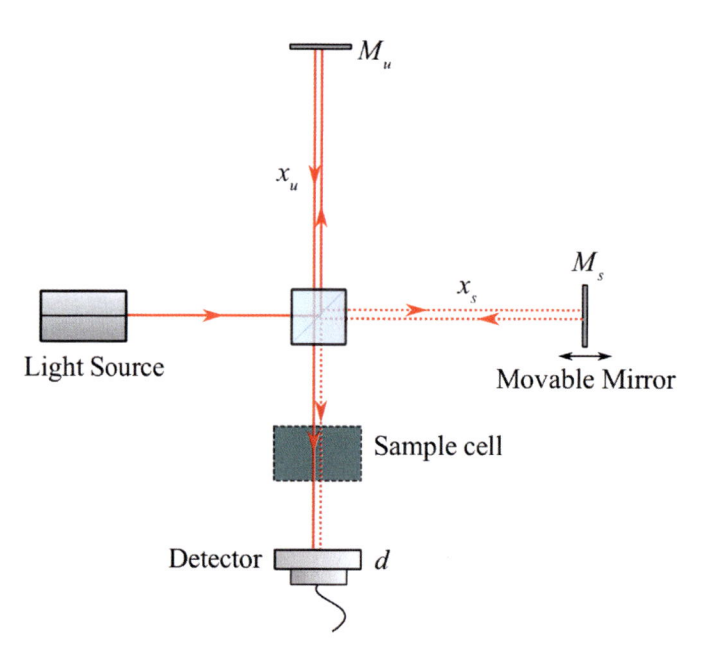

Figure 8.4 – Michelson interferometer with a movable mirror. With the intensity recorded by the detector as a function of $\xi = 2(x_s - x_u)$, the spectrum of the light can be determined from the Fourier transform of the recorded signal. By adding a sample cell before the detector the absorption spectrum of the material within the cell can be determined in the same manner.

The distance between the beam splitter and M_u is x_u, and that between the beam splitter and M_s is x_s. Light from a source reflects off the beam splitter for a round trip to and from M_u of $2x_u$, while the transmitted light traveling through the beam splitter horizontally undergoes a round trip to and from M_s of $2x_s$. The beam reflected off M_u transmits downward through the beam splitter and that reflected from M_s next reflects against the beam splitter and also heads downward. These two beams interfere at the detector d. A spectrum is taken by continuously moving mirror M_s over a known displacement while recording the interferogram. In what follows we will show that the interferogram is associated with the inverse Fourier transform of the spectrum of the light source.

The two waves leading to interference at the detector d with field E_u and E_s, with the former associated with reflection off M_u and the latter with reflection off M_s generate

$$E_d = E_u + E_s = E_0 e^{ik2x_u}\left[1 + e^{ik(2x_s - 2x_u)}\right], \quad (8.8)$$

where we have assumed that the two waves approaching the detector d have polarizations perpendicular to the page and are of equal amplitudes E_0.

The detector measures intensity, which will be represented as $E_d^* \cdot E_d$,

$$E_d^* E_d = E_0^2\left[2 + e^{ik(2x_s - 2x_u)} + e^{-ik(2x_s - 2x_u)}\right] = 2E_0^2\left[1 + \frac{e^{ik\xi} + e^{-ik\xi}}{2}\right] \quad (8.9)$$

where the round trip difference in path lengths $2(x_u - x_s)$ is ξ. On this basis the detected intensity is

$$I_d(\xi) = \frac{I_d(0)}{2}\left[1 + \frac{e^{ik\xi} + e^{-ik\xi}}{2}\right], \quad (8.10)$$

where $I_d(0)$ is the intensity at equal path lengths, and $I_d(0)/2$ is the average intensity with respect to ξ. An experimenter using monochromatic light sees the intensity varying as $I_d(0)\cos^2(k\xi/2)$ as the side mirror is moved. This allows the wavelength of the laser to be determined. If it takes a motion Δx_s of the side mirror to produce one period of oscillation in I_d, also known as a "fringe", then $k = \pi/\Delta x_s$; from which the wavelength of the laser $\lambda = 2\Delta x_s$. Refractive index can also be measured if a gas cell is put in one leg of the interferometer and the gas is pumped out while observing the number of fringes that appear in I_d (see Ex.8.3). There is no need to move the mirror; what is changing is the refractive index within the cell, and that changes k in the cell, since $k = nk_0$. To do the spectroscopy of molecules in a liquid will require that we change the light source.

Now instead of a monochromatic source we consider a broadband radiation filtered by the transmission through an unknown sample with a spectral density $B(k)$ such that the intensity at equal arms length $I_d(0)$ is given by

$$\frac{I_d(0)}{2} = \int_0^\infty B(k)\,dk. \tag{8.11}$$

As the side mirror is moved each differential spectral $B(k)dk$ component is intereference modulated as in Eq. (8.10), so that the detected intensity

$$I_d = \int_0^\infty B(k)\left[1 + \frac{e^{ik\xi} + e^{-ik\xi}}{2}\right]dk = \frac{I_d(0)}{2} + \int_0^\infty B(k)\left[\frac{e^{ik\xi} + e^{-ik\xi}}{2}\right]dk. \tag{8.12}$$

It is possible to transform the integral on the right of Eq. (8.11) into an inverse Fourier transform by allowing $B(-k) = B(k)$;

$$I_d(\xi) = \frac{I_d(0)}{2} + \frac{1}{2}\int_{-\infty}^\infty B(k)e^{ik\xi}\,dk \quad \text{and} \tag{8.13}$$

$$A(\xi) = \frac{2I_d(\xi) - I_d(0)}{2\pi} = \frac{1}{2\pi}\int_{-\infty}^\infty B(k)e^{ik\xi}\,dk. \tag{8.14}$$

$A(\xi)$ readjusts the baseline of the signal $I_d(\xi)$ by subtracting off the constant portion $I_d(0)/2$, and thereby relates it to the inverse Fourier transform of the spectrum $B(k)$ by dividing by π. The Fourier transform of $A(\xi)$ is

$$B(k) = \int_{-\infty}^\infty A(\xi)\cdot e^{-ik\xi}\,d\xi, \tag{8.15}$$

which allows the spectrum to be determined. Clearly $A(\xi)$ is the most important part of the detected intensity, and for that reason we will assign it the name *detected signal*.

Fig. 8.4a shows an example of the relation between the spectral density $B(k)$ and the detected signal $A(\xi)$. On the left $B(k)$ has a Gaussian form (see Ex.8.5) with a peak in spectral brightness at a wavelength $\lambda_{max} = 0.628\,\mu m$, i.e. $k = 10^7\,m^{-1}$, and a width in wavelength between 1/e points of 250nm, i.e. $(\Delta k)_{1/e} = 0.4 \times 10^7\,m^{-1}$. On the right is the associated detected signal which falls off rapidly from the point of equal path lengths (i.e. $\xi = 0$). Here we see a fringe pattern with a spatial period in ξ of λ_{max}. $A(\xi)$ is clearly modulated by a Gaussian envelope.

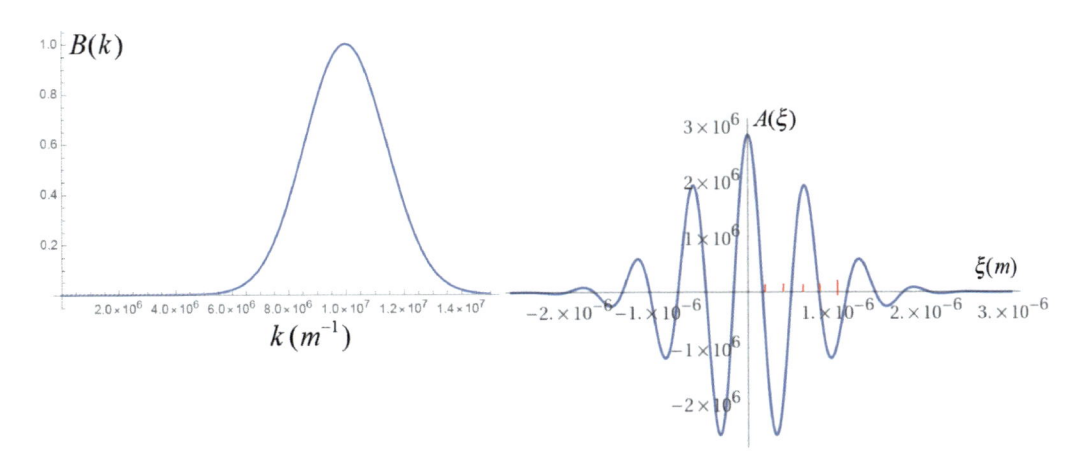

Fig. 8.4 – On the left is shown the spectral density $B(k)$ for a possible light source, and on the right is shown the detected signal $A(\xi)$. Note that the separation between fringes in terms of the change of round trip path difference on the right is approximately the average wavelength in the spectrum on the left.

If we characterize the full width of $B(k)$ at 1/e of the maximum by $(\Delta k)_{1/e}$ and the associated width of the $A(\xi)$ envelope by $(\Delta \xi)_{1/e}$, then one can show that the product of the two widths is about 8;

$$(\Delta \xi)_{1/e}(\Delta k)_{1/e} \approx 8. \tag{8.16}$$

So a light source that is spectrally broad will have few apparent fringes, with these fringes existing near equal arms length where the signal $A(\xi)$ is maximal. For our example source $(\Delta \xi)_{1/e}$ is about $2\mu m$, however if we had used the solar spectrum which is about 4× wider, then $(\Delta\xi)_{1/e}$ would be about $0.5\mu m$. On the other hand a laser which is ideally monochromatic will have in principle a huge number fringes with equal arms length not distinguished from a round trip displacement many wavelengths away.

It is important at this point to describe the use of the interferometer for spectroscopy. The side mirror is moved symmetrically so that the span covers the region from $\xi = -L$ to $\xi = +L$, and the detected intensity $I_d(\xi)$ is recorded. This signal is converted to $A(\xi)$ as in Eq. (8.14), and the spectrum $B(k)$ is determined from the Fourier transform of $A(\xi)$. That is all there is to FTIR. As in all spectrometers resolution is of great importance. What does it depend on for FTIR?

The fact that a Fourier transform connects the absorption spectrum to the signal allows us to connect up the spatial frequency line width of the feature in the intensity spectrum $(\delta k)_{1/2}$ to the length over which the mirror is moved. Evaluating Eq. (8.15) over a truncated length $2|L|$ causes a broadening of features by

$$(\delta k)_{1/2} = \frac{\pi}{L}. \qquad (8.17)$$

So the longer distance the mirror is moved, the better the resolution, the linewidth Eq. (8.17) is a statement about the minimum separation in frequency since $k = \omega/c$. FTIR spectroscopists prefer to work with "wave number" $1/\lambda$, so that Eq. (8.17) reads $(\delta(1/\lambda))_{1/2} = 1/(2L)$. As an example of this, let's say that we would like to separate the prominent yellow D-lines of sodium. These lines are at 589.0 nm and 589.6 nm. Our rules then says that the minimum length that can be scanned to resolve those lines separately is

$$L = \frac{1}{2\left[\dfrac{1}{\lambda_1} - \dfrac{1}{\lambda_2}\right]}. \qquad (8.18)$$

Surprisingly, this length is only 289.4 micrometers, so the precision with which the movement of the mirror must be recorded requires extremely good metrology.

Fig. 8.5 provides an example of what can be learned from FTIR about a sample installed within the sample cell in Fig. 8.4. The dips in the figure show the wavelengths of light that are absorbed by the polyethylene molecules.

In general, FTIR spectroscopy measures the spectrum of light that is absorbed by a sample. Analyzing the frequencies that are absorbed helps determine chemical characteristics, such as vibrational energy modes, of a sample.

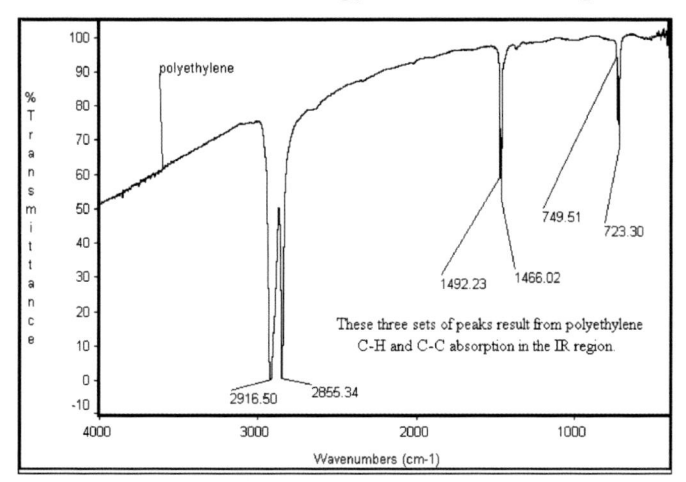

Figure 8.5 – Spectrum of Polyethylene determined using FTIR.

The following table summarizes the resolution of each of the spectrometers discussed in this chapter.

Instrument	Wavelength Resolution	Reference
Grating Spectrometer (1st order) N = "slits" illuminated	$\Delta \lambda = \dfrac{\lambda}{N}$	Sec.8.1, Eq. (8.7)
FTIR Spectrometer L = full mirror displacement in scan	$\Delta\left(\dfrac{1}{\lambda}\right) = \dfrac{1}{2L}$	Sec.8.2, Eq. (8.18)

Resolution may be thought of as the amount by which a very narrow spectral line is broadened by the particular instrument. Aside from instrument broadening, most light sources are naturally broadened through the emission process.

8.3 – Chapter Eight Exercises

Exercise 8.1 – Fig. 8.3 shows the line width of the 0th order diffraction spot as broader than the 1st order spot. Is that real or a misrepresentation? Explain your reasoning.

Exercise 8.2 – What does the detector output of a Michelson FTIR spectrometer $I_d(\xi)$ look like if the light source consists of the two sodium D lines, and the round trip difference in path lengths ξ varies continuously from -2 mm to +2 mm? Hint: express the input intensity spectrum $B(k) = \delta(k - k_1) + \delta(k - k_2)$, and take the associated wavelength for the two sources to be 589.0 nm and 589.6 nm.

Exercise 8.3 – An airtight chamber 5.0 cm long with glass windows is placed in one arm of a Michelson interferometer as indicated in the figure below. Light of the wavelength $\lambda = 500 nm$ is used. The air is slowly evacuated from the chamber using a vacuum pump. While the air is being removed, 60 fringes are observed to pass through the view. From these data, find the index of refraction of air at atmospheric pressure.

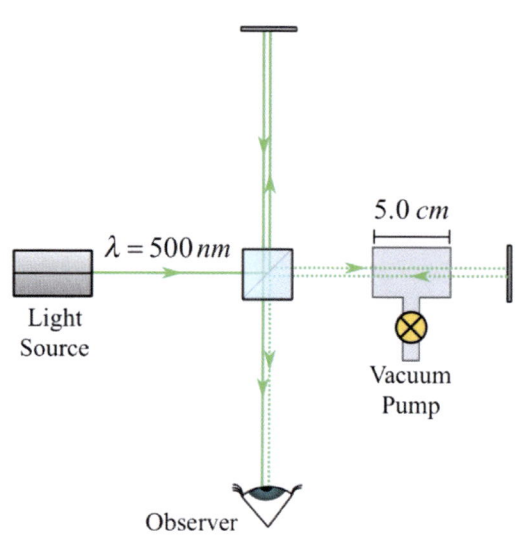

Exercise 8.4- Consider the Michelson spectrometer in Fig. 8.4, and Eq. (8.14). By writing the $B(k)$ as a single sided function $B_+(k)$ [i.e. $B_+(k) = B(k)$ for $k > 0$, but $B_+(k) = 0$ for $k < 0$], show that the output from the Michelson spectrometer in Fig. 8.4 is

$$A(\xi) = 2 \, \text{Re}\{F^{-1}[B_+(k)]\},$$

where F^{-1} is the inverse Fourier Transform.

Exercise 8.5 – Using the result of Exercise 8.4, find the output of a the Michelson interferometer for a light source having a Gaussian spectral density,

$$B_+(k) = \exp\{-[(k - 10^7 \, m^{-1})/0.2 \times 10^7 \, m^{-1}]^2\}.$$

186

Chapter Nine – Geometrical Optics

9.1 – Fermat's Principle of Least Time

Geometrical optics concerns itself with the description of how light is manipulated by large-scale objects (i.e. objects much larger than a wavelength). It is sometimes referred to as ray optics. The amazing thing is that there is only one basic idea that governs how rays behave; it is Fermat's least time principle. In short the principle states that light travels over a path that takes the least time between two points a and b regardless of the inhomogeneity of the medium in between them. Since light travels a distance ds in a time $n(s)ds/c$, the overall time to move from a to b is

$$t_{a \to b} = \frac{1}{c}\int_a^b n(s)\,ds. \tag{9.1}$$

The integral over the refractive index along the path is called the "optical path length" (OPL). It is this length that needs to be minimized. The integral is depicted in Fig. 9.1.

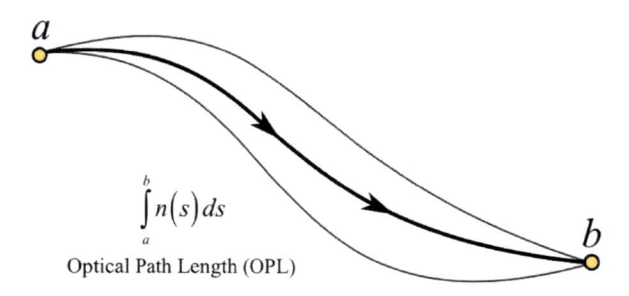

Figure 9.1 – Fermat Integral.

Using Fermat's principle is a convenient way to arrive at many rules for determining the path that light takes. To do this one simply takes an equation for the OPL with respect to a parameter and finds where the derivative of the OPL equation with respect to the parameter is zero, i.e. the OPL is minimum. We will now use this principle to arrive at some familiar rules and then extend it to deal with lenses and materials with varying refractive indices.

Fermat's Principle applied to reflection and refraction

As a 1st example least time in a homogeneous medium means traveling in a straight line. This is the simplest rule in geometrical optics, and is explained simply by Fermat's principle.

As a 2nd example light is reflected from a mirror as shown in Fig. 9.2. By minimizing the time for light to be launched from a point a and signal a point b (Fig. 9.2A) we obtain the law of reflection $\theta_i = \theta_r$ (Fig. 9.2B); a second separate rule we learned in geometrical optics.

(A)

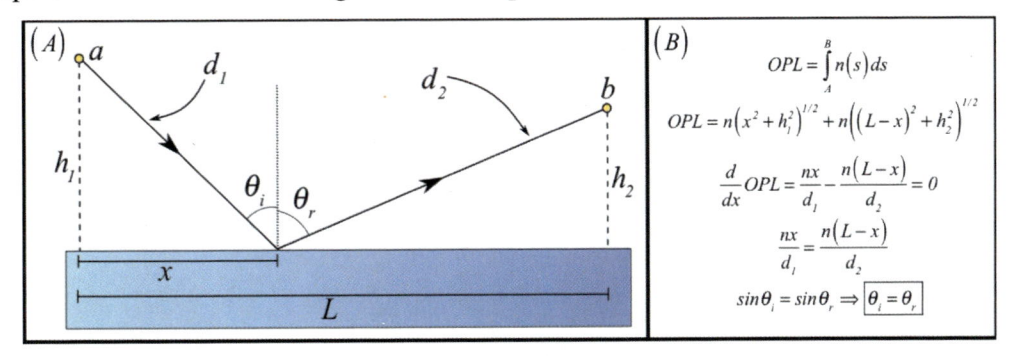

(B)
$$OPL = \int_A^B n(s)\,ds$$
$$OPL = n(x^2 + h_1^2)^{1/2} + n((L-x)^2 + h_2^2)^{1/2}$$
$$\frac{d}{dx}OPL = \frac{nx}{d_1} - \frac{n(L-x)}{d_2} = 0$$
$$\frac{nx}{d_1} = \frac{n(L-x)}{d_2}$$
$$\sin\theta_i = \sin\theta_r \Rightarrow \boxed{\theta_i = \theta_r}$$

Figure 9.2 – Fermat's Principle applied to reflection at a surface. (A) Diagram of the general path taken by light reflecting off of a flat surface, x can be varied to change the path length. (B) The law of reflection can be derived by minimizing the optical path length with respect to x.

As a 3rd example we look at the case of the light penetrating a surface i.e. from air to water (Fig. 9.3A). In this case the Optical Path Length (OPL) will involve two refractive indices (n_1 for air and n_2 for water). The result derived in Fig. 9.3B is Snell's Law $n_1 \sin(\theta_1) = n_2 \sin(\theta_2)$; another separate rule from geometrical optics is again explained using Fermat's least time principle.

(A)

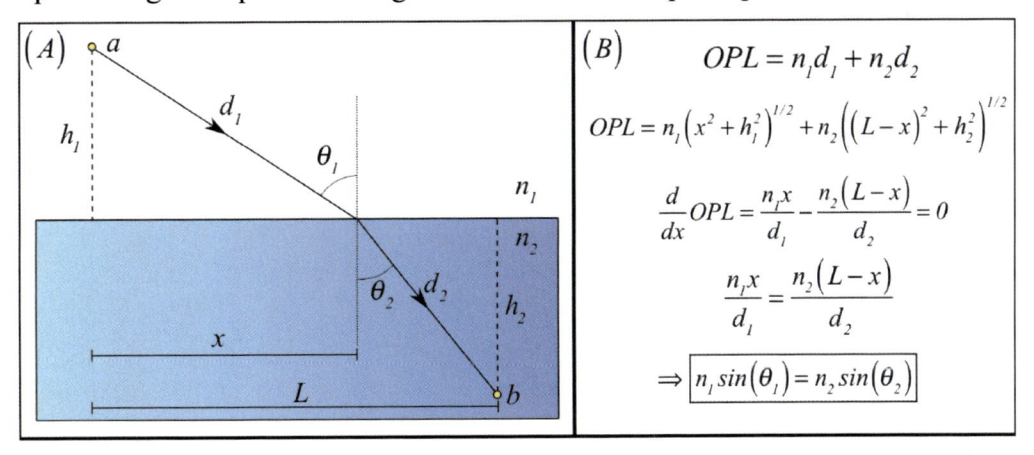

(B)
$$OPL = n_1 d_1 + n_2 d_2$$
$$OPL = n_1(x^2 + h_1^2)^{1/2} + n_2((L-x)^2 + h_2^2)^{1/2}$$
$$\frac{d}{dx}OPL = \frac{n_1 x}{d_1} - \frac{n_2(L-x)}{d_2} = 0$$
$$\frac{n_1 x}{d_1} = \frac{n_2(L-x)}{d_2}$$
$$\Rightarrow \boxed{n_1 \sin(\theta_1) = n_2 \sin(\theta_2)}$$

Figure 9.3 – Fermat's Principle applied to refraction through a surface. (A) Light is bent at the interface and the path length varies with x. (B) Snell's law is found by minimizing the optical path length with respect to x.

Fermat's Principle Applied to the Hubble Space Telescope Mirror (4)

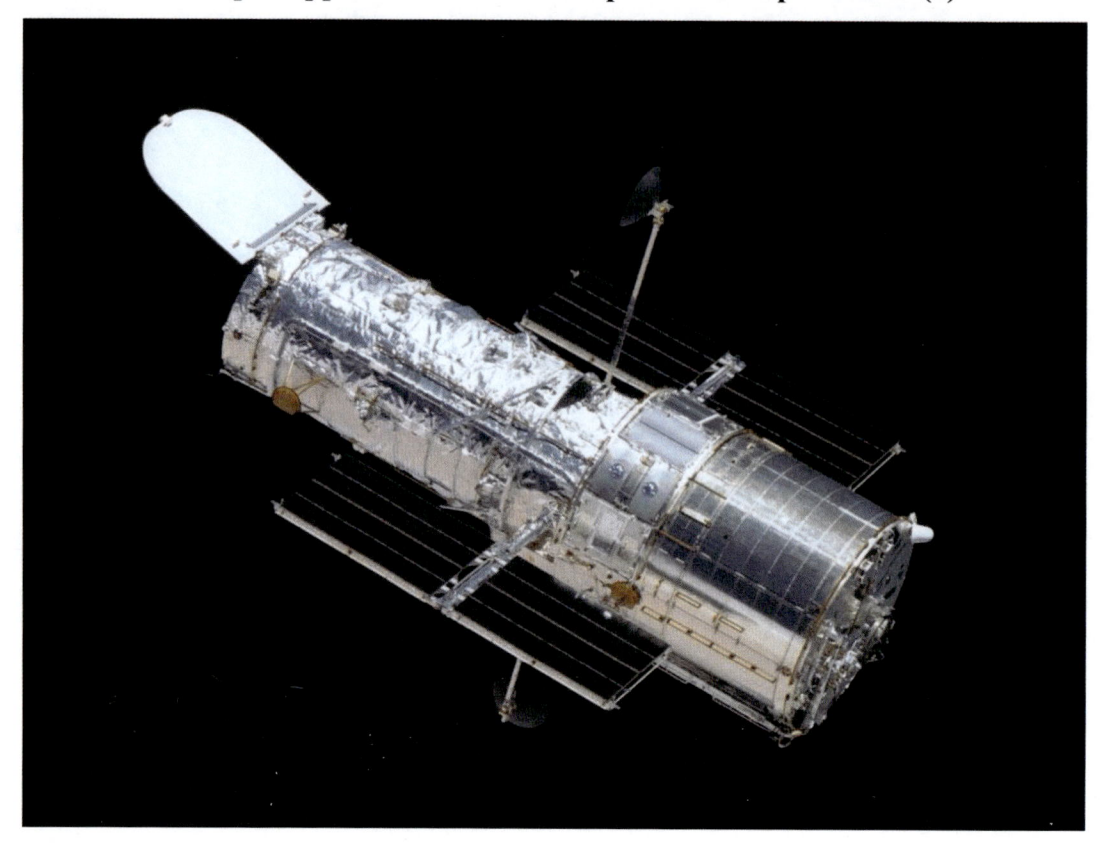

Figure 9.4 – The Hubble Telescope. The mirror used to focus light in the telescope is 8 ft. in diameter. (Photo credit: NASA).

As a 4th example we consider a curved mirror like the one in the Hubble telescope (Fig. 9.4) and wonder what shape can bring all parallel rays of light from a distant star to a single point, the point of focus (Fig. 9.5). The analysis is a bit tricky. For all rays to get to the same point they must travel the same OPL. The ray on the centerline as it moves beyond the horizontal level of the focal point travels an overall distance $2f$ to get back again. So all rays moving parallel to the axis of symmetry upon crossing this same horizontal level, shown in Fig. 9.5, must travel that same distance in getting to the focal point. The ray on the right travels a distance $f - y$ to get to the mirror, and must travel an additional distance d to get to the focus. However $d^2 = (f - y)^2 + x^2$. So

$$2f = (f-y) + \left[(f-y)^2 + x^2\right]^{1/2}. \tag{9.2}$$

A little more algebra and we arrive at the shape $y = x^2/(4f)$, a parabola, which is in fact the shape of the Hubble satellite's mirror.

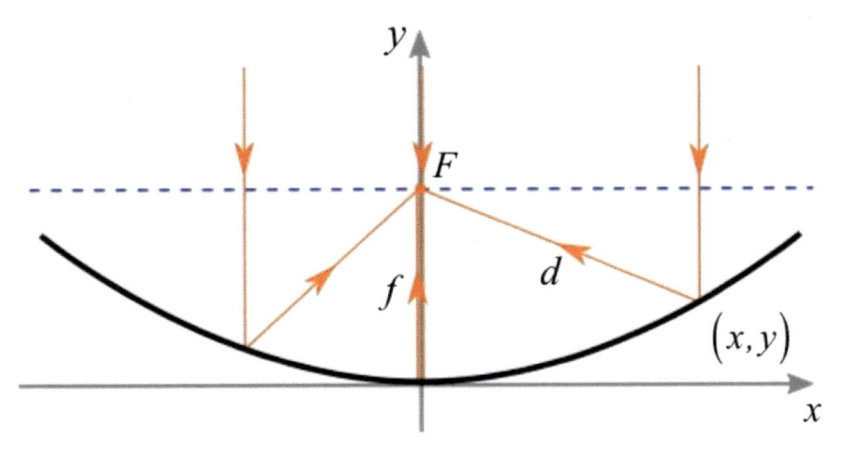

Figure 9.5 – What is the shape of a mirror that can focus all rays directed along its symmetry axis at a point?

Spherical mirrors are often substituted for parabolic mirrors because they are far less expensive to manufacture. Near the bottom of a parabola a circle comes very close to the same shape as a parabola (Fig. 9.6).

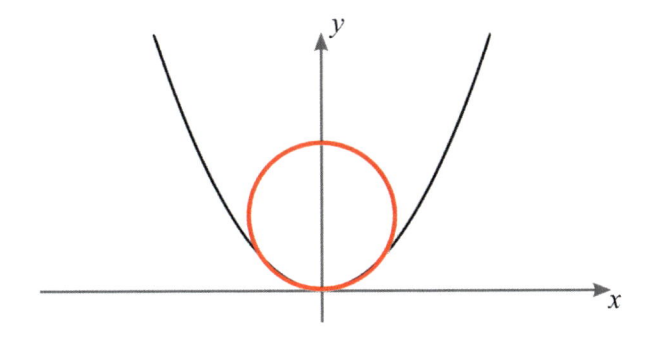

Figure 9.6 – A circle compared with a parabola.

To see how different the circle is from a parabola (Fig. 9.7), we need to make some approximations.

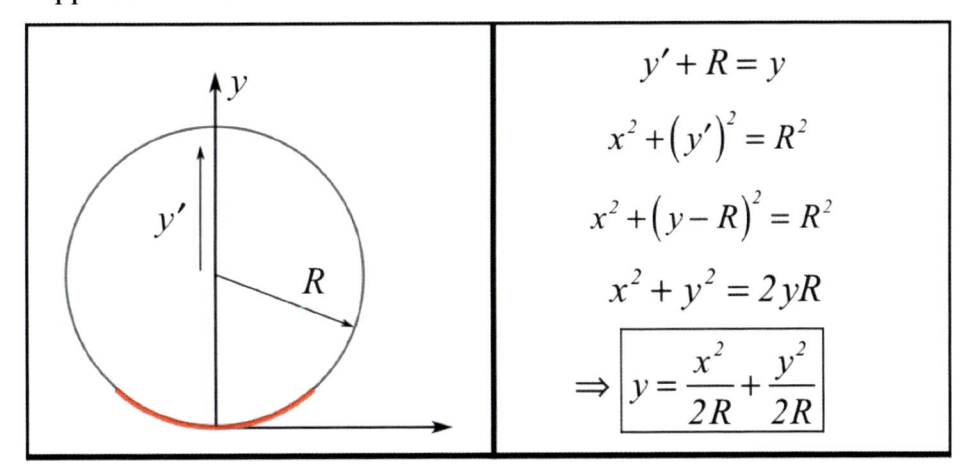

$$y' + R = y$$
$$x^2 + (y')^2 = R^2$$
$$x^2 + (y - R)^2 = R^2$$
$$x^2 + y^2 = 2yR$$
$$\Rightarrow \boxed{y = \frac{x^2}{2R} + \frac{y^2}{2R}}$$

Figure 9.7 – How close is the shape of the bottom of a circle to a parabola?

With the y-coordinate starting from the bottom of the circle, it resembles a parabola for $|x| \ll R$, in as much as $y \approx x^2/(2R)$ in this limit, since y increases much more slowly than x. The first correction to this can be arrived at by a successive iteration. By putting our asymptotic solution back into the exact formula (Fig. 9.6 on the right) we get $y \approx x^2/(2R) + x^4/(8R^3)$, which is equal to $y \approx x^2/(2R)\left[1 + x^2/(4R^2)\right]$. So if x is within $R/8$ we deviate from a parabola by about 1 part in 256, a pretty good approximation to a parabola. Since the parabola has a perfect focus with the equation $y = x^2/(4f)$, a mirror for which $|x| \ll R$ should have the same focus with the equation $y \approx x^2/(2R)$, which suggests that $4f$ for the mirror is $2R$, from which the focal length of the asymptotic mirror is $R/2$. By using this along with the definition of the focal length, the mirror imaging equation is easily arrived at (Fig. 9.8),

$$\frac{1}{s_o} + \frac{1}{s_i} = \frac{1}{f} = \frac{2}{R}. \qquad (9.3)$$

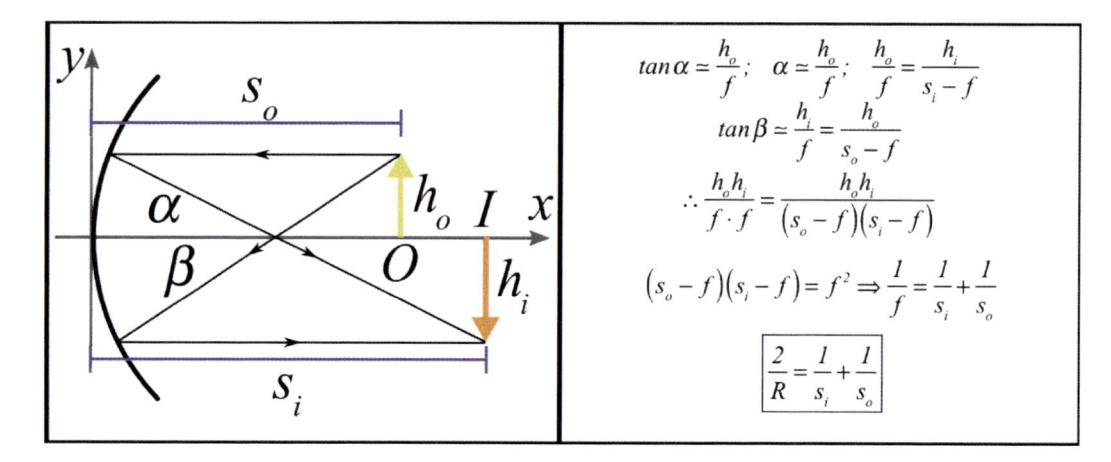

Figure 9.8 – Spherical mirror imaging equation.

This is a good time to introduce the transverse magnification M_T. It is the ratio of the image height to the object height, and carries a sign; positive if the image is upright, and negative if it is inverted. It can be related to the image and object distances from the mirror,

$$M_T = \frac{h_i}{h_o} = -\frac{s_i}{s_o}. \qquad (9.4)$$

Fermat's Principle applied to imaging by a spherical refracting surface

As a 5th example we consider refraction and image formation through a spherical dielectric surface. This is a problem that is very dear to me. My introduction to this subject was experimental. At an early age I was using grinding powder to shape a chunk of glass at the Planetarium. My first job was to grind a spherical surface from a chunk of glass with the eventual goal of producing a biconvex lens. The mathematics to describe how the spherical surface refracts light was not taught to me at that time. I learned the fundamentals later in college. The approach is to wonder how a ray of light emanating from an object on the symmetry axis of a lens will behave when it strikes a spherical surface (note the point on the symmetry axis to the left of the spherical surface in Fig. 9.9A).

(A) [diagram of spherical refracting surface with n_1, n_2, ℓ_o, ℓ_i, R, ϕ, $\pi-\phi$, O, I, C, s_o, s_i]

(B)
$$\ell_o^2 = R^2 + (s_o + R)^2 - 2R(s_o + R)\cos\phi$$
$$\ell_i^2 = R^2 + (s_i - R)^2 + 2R(s_i - R)\cos\phi$$
$$OPL = n_1 \ell_o + n_2 \ell_i$$
$$\frac{\partial OPL}{\partial \phi} = 0$$
$$\Rightarrow \frac{n_1}{\ell_o} + \frac{n_2}{\ell_i} = \frac{1}{R}\left(\frac{n_2 s_i}{\ell_i} - \frac{n_1 s_o}{\ell_o}\right)$$

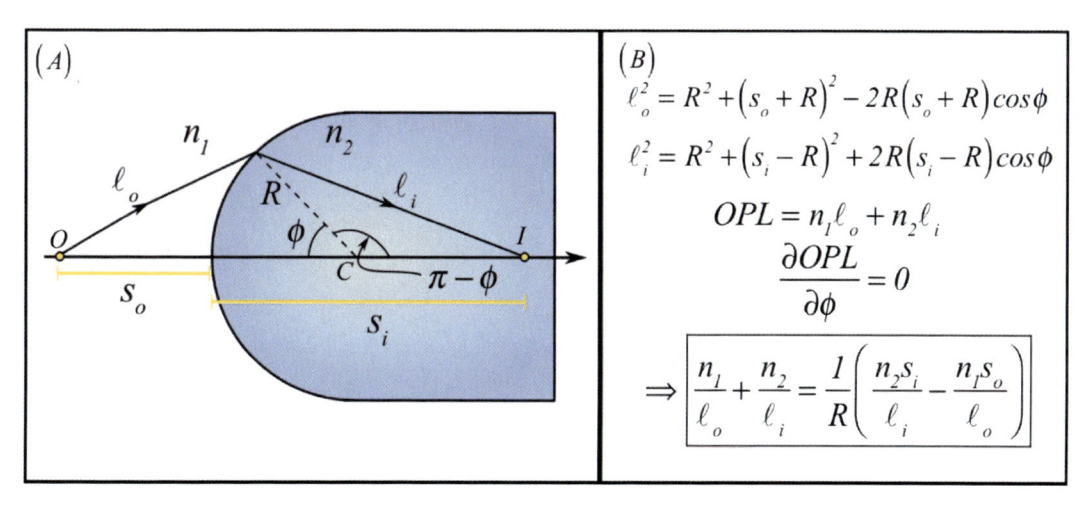

Figure 9.9 – Imaging through a spherical refracting surface. (A) The path of light refracting from a spherical surface can be varied with parameter ϕ. (B) Minimizing the OPL with respect to ϕ gives the path taken by light.

The refractive index on the left side is n_1 and that on the right is n_2. Fig. 9.9B shows the analysis based on Fermat's least time principle. Only one variable ϕ is used for describing the optical path length (OPL), and by executing the least time principle, $\frac{d(OPL)}{d\phi} = 0$, we obtain

$$\frac{n_1}{\ell_o} + \frac{n_2}{\ell_i} = \frac{1}{R}\left(\frac{n_2 s_i}{\ell_i} - \frac{n_1 s_o}{\ell_o}\right). \qquad (9.5)$$

Eq. (9.5) is exact and shows a dependence on not only the object and images distances (s_o, s_i), but also the distance the light travels to the surface ℓ_o, as well as the distance from the surface to the symmetry axis ℓ_i. This is because a spherical refracting surface cannot produce the same image distance independent of where the ray strikes the surface. It is only in the limit of paraxial rays that one image is produced. This is what is commonly known as spherical aberration. In the limit of small ϕ the $\cos(\phi) \approx 1$, and ℓ_o and ℓ_i become s_o and s_i, respectively, and the single surface imaging equation reduces to

$$\frac{n_1}{s_o} + \frac{n_2}{s_i} = \frac{(n_2 - n_1)}{R}. \tag{9.6}$$

There are important sign conventions that need to be followed in using Eq. (9.6):

- s_o is positive only in real object space (the space from which the object ray is launched, before hitting the surface, e.g. to the left of the interface in Fig. 9.9).

- s_i is positive in real image space (the space in which film placed at the image would photograph it, e.g. to the right of the interface in Fig. 9.9).

- R is positive with the center of curvature C in real image space, and negative otherwise.

A great deal can be learned from Eq. (9.6) about problems for which it is appropriate. For example arguments about the actual and apparent depth of a swimming pool to an outside observer can be rationalized.

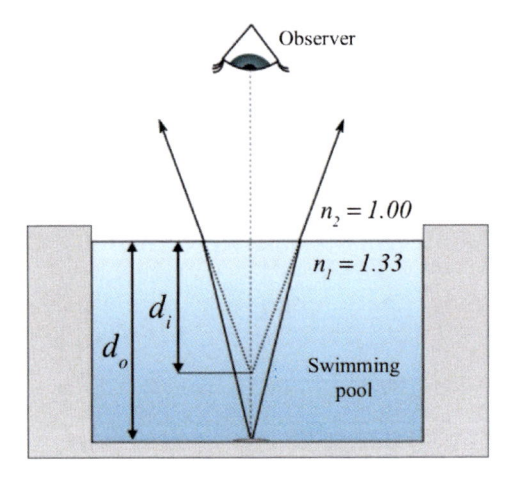

Figure 9.10 – The swimming pool problem. Why does the pool appear to be shallower than advertised?

In particular the owner of the pool says with pride that it is 6 feet deep, however a viewer looking directly down at the swimming pool drain has the impression that it is shallower. Fig. 9.10 illustrates why the apparent depth d_i can never be the same as the actual depth d_o. To calculate how far these are apart we set $R = \infty$ in Eq. (9.6) for a flat surface, and find that $s_i = -(n_2/n_1)s_o$, with $n_1 = 1.33$, $n_2 = 1$ and $s_o = 6\,ft$, we find that $s_i = -4.5\,ft$. This is clearly shallower with the minus sign signifying a virtual image in the water.

Fermat's Principle Applied to Graded Index Glass (6)

Common glass is made of silica (SiO_2) that has a spatially uniform refractive index. So light travels in a straight line in bulk. However by adding a Germanium dopant (GeO_2) with a gradient of density, the local refractive index is increased in proportion to the concentration of (GeO_2). That's basically because the Germanium is more easily polarized by light than Silica; Germanium has a larger polarizability α (i.e. the dipole moment $\mu = \alpha E$, where E is the field). Doping glass in this way leads to a means for guiding light. To understand the concept it is best to look at a graded index fiber in which the refractive index is highest in the center and falls of parabolically away from center. Fig. 9.11 shows what happens. Light moving from the center while traveling from left to right curves along its light path. To understand how this works we break up the radial refractive index profile into steps of varying refractive index. As the light moves outward Snell's "law" which was arrived at from Fresnel's principle dictates that if the next interface has a lower refractive index then it must increase the angle of refraction. This goes on as shown until the condition for total internal reflection is reached at which point the light reflects downward. On its way down this light encounters a step with higher refractive index where Snell's law demands that it move down more steeply. The net effect is that the rays follow a sinusoidal orbit. You might wonder why this costly approach is followed, since a simple step index fiber (i.e. A fiber in which the core has uniform refractive index) could be used. The problem with a step index fiber is that the time to go from the launch point to the detector is spread out depending on the launch angle. Fortunately the graded index approach taken in Fig. 9.11 has far less dispersion of the transit time from source to detector.

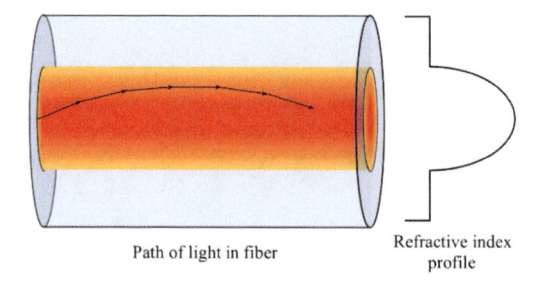

Figure 9.11 – Fiber in which the core has a radially varying refractive index profile.

9.2 – Constructing a Lens

Now I'm brought back to the Planetarium story. Once having ground the front side (left) of the chunk of glass which took about two weeks of intensive labor, I was asked to grind the other side, with the motivation being that we were going to build a refracting telescope, and the major lens would be my handiwork. This introduces the next step in my analysis, the lens. I will lean heavily on the paraxial result (Eq. (9.6)) as we move forward.

$$\frac{n_1}{s_{o1}} + \frac{n_2}{s_{i1}} = \frac{n_2 - n_1}{R_1}$$

$$\frac{n_2}{-(s_{i1}-t)} + \frac{n_1}{s_{i2}} = \frac{n_1 - n_2}{R_2}$$

For small t adding these equations gives

$$\frac{1}{f_{TL}} = \frac{(n_2 - n_1)}{n_1}\left(\frac{1}{R_1} - \frac{1}{R_2}\right)$$

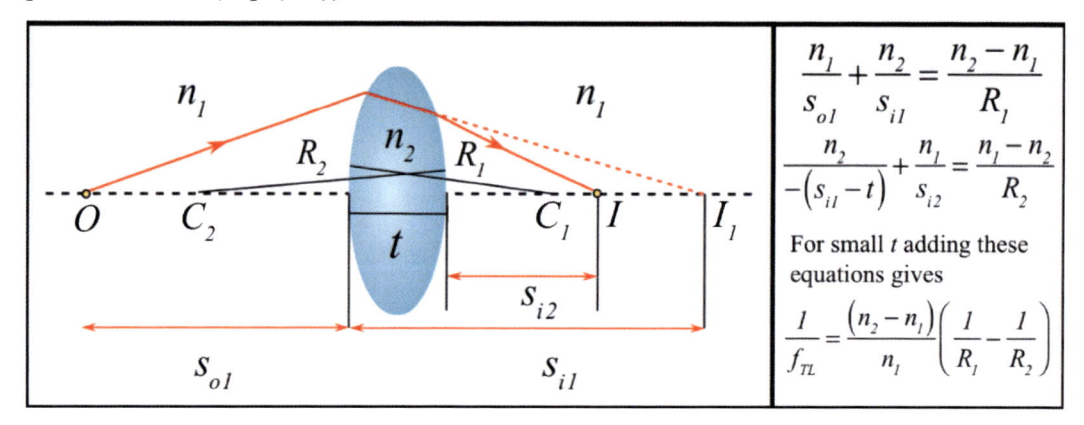

Figure 9.12 – A lens that has been ground on both sides.

Fig. 9.12 shows the geometry of a typical lens, for imaging an object radiating at s_{o1}. The ray trace in red shows the entire path of the light to the corresponding image point at s_{i2}. Had there been no second surface the ray after hitting the first surface would have proceeded to s_{i1}. The first equation on the right in the figure contains the analysis for this trajectory based on Eq. (9.6). As far as the second surface is concerned the light is heading for s_{i1}, so the image formed by the first surface is the virtual object for the second surface, and at a distance $s_{i1} - t$ from the second surface. However the object is virtual, and by our sign convention, should have a negative object distance, $-(s_{i1}-t)$, as indicated in the second equation in Fig. 9.12. You will also note that the refractive index associated with this object is that of the glass, n_2, since the object forms as if it were in the glass. On the other hand the image of surface 2 at s_{i2} forms in air, or whatever is the surrounding medium. The reversal of the refractive index difference is opposite of what it is for the first surface, consistent with the refractive indices used on the left of the second equation. Combining the two equations gives

$$\frac{1}{s_{o1}}+\frac{1}{s_{i2}}=\frac{(n_2-n_1)}{n_1}\left(\frac{1}{R_1}-\frac{1}{R_2}\right)+\frac{n_2}{n_1}\left(\frac{t}{s_{i1}(s_{i1}-t)}\right). \qquad (9.7)$$

The effect of the thickness of the lens on the imaging process is contained in the last term on the right in Eq. (9.7). It is convenient and appropriate to look at the special case of a thin lens when the thickness t is much smaller than all other lengths in this equation. Under the thin less approximation t is taken to be very small and the term containing t is set to zero. Under this approximation it is meaningless to reference the object distance to the first surface and the image distance to the second. Rather they are both reference to the position of the lens with s_{o1} and s_{i2} set equal to s_o and s_i respectively, and the equation that results is called the thin lens equation;

$$\frac{1}{s_o}+\frac{1}{s_i}=\frac{(n_2-n_1)}{n_1}\left(\frac{1}{R_1}-\frac{1}{R_2}\right). \qquad (9.8)$$

By setting the object distance to infinity, the image is formed at f, the back focal point of the lens, as in Fig. 9.13.

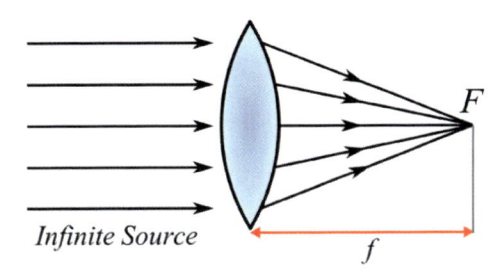

Figure 9.13 – The focal length, f, is the distance from lens at which light rays exiting the lens meet at a point.

Eq. (9.8) is then rewritten as

$$\frac{1}{s_o}+\frac{1}{s_i}=\frac{1}{f} \text{ with} \qquad (9.9)$$

$$\frac{1}{f}=\frac{(n_2-n_1)}{n_1}\left(\frac{1}{R_1}-\frac{1}{R_2}\right). \qquad (9.10)$$

Eq. (9.10) is known as the lens maker's equation since it enables the design of the most important parameter in a thin lens, the focal distance f. It should be noted that inverse focal length $1/f$ has a great deal of significance in optometry and is given the name Power with the special unit Diopter (D) when focal length is expressed in meters. Depending on the relative sizes of R_1 and R_2, f can be positive or negative. In the next section we will look at two-lens systems and see why lens power is so important. Before this however we observe how the lens makers equation works for lenses meant to be used as spectacles. Fig. 9.14 shows some rather powerful spectacle lenses. They are typically concave or perhaps flat on the side facing the eye and convex on the outside. Although all have this feature in common their powers can be positive or negative.

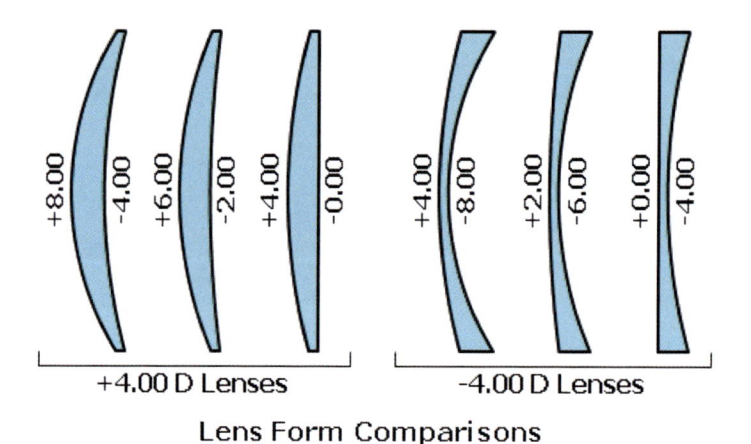

Figure 9.14 – A variety of crescent (concave out, concave in) shaped lenses for correcting both farsightedness (left) and nearsightedness (right).

Note that there appear to be numbers associated with each surface of the lenses. These numbers are the contributions to the overall power of the lens. In each case the power of the front surface in air, is $D_1=(n-1)/R_1$, and the power of the back surface, is $D_2=(1-n)/R_2$, with the overall power $P=D_1+D_2$, consistent with Eq. (9.10). Note for this particular type of lens [i.e. (convex out, concave in)] both R_1 and R_2 are positive numbers, by our convention (i.e. both centers of curvature are on the right), so that the reversal of sign between D_1 and D_2 is associate with the refractive index change as one crosses the surface from left to right.

When the thin lens equation is no longer accurate enough due to significant relative thickness, a new approach to calculation and new notation must be used. The ray trace becomes a bit more complicated as shown in Fig. 9.15. Here an effective focal length f_{eff} replaces the thin lens focal length in Fig. 9.12, and the rays are depicted as refracting at principle planes H_1 and H_2. Upon these planes the exiting ray from the lens meets the incident ray. Note that the effective focal length is measured from a principal plane to point of focus. A key thing is that the back and front focal lengths are no longer equal as they were for the thin lens.

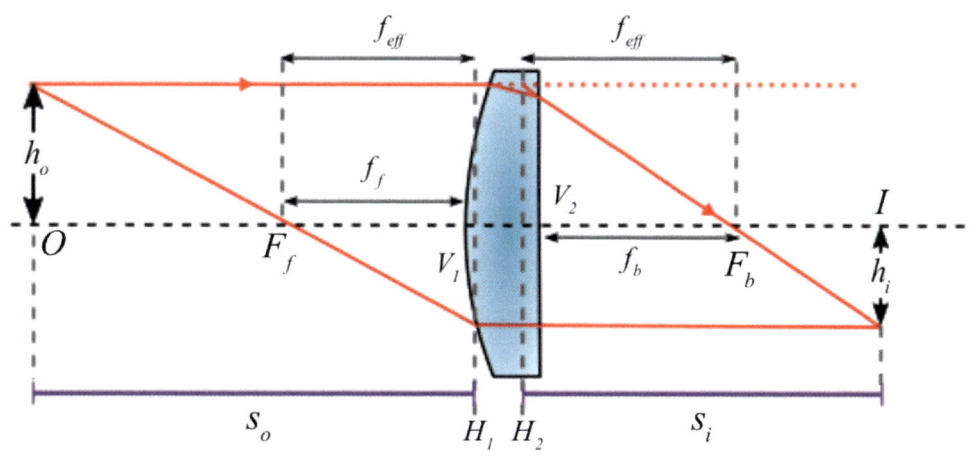

Figure 9.15 – Ray tracing through a thick lens.

We will delve deeper into the thick lens in Section 9.4 on matrix optics, however for now it is important to understand that the form of the imaging equation

$$\frac{1}{s_o} + \frac{1}{s_i} = \frac{1}{f_{eff}}, \qquad (9.11)$$

still persists, however the thin lens focal length f in Eq. (9.10) is replaced by f_{eff}.

Certainly, the lens maker's equation is changed from Eq. (9.10)

$$\frac{1}{f_{eff}} = (n_l - 1)\left(\frac{1}{R_1} - \frac{1}{R_2} + \frac{(n_l - 1)t}{n_l R_1 R_2}\right), \qquad (9.12)$$

however, one can see Eq. (9.12) is consistent with Eq. (9.10) as the lens becomes thin and t vanishes. Note that when the thickness t becomes negligibly small by comparison with the lens radii, Eq. (9.12) returns to the thin lens equation (Eq.

199

(9.10)). After the next section, which deals with the focal length for a two-lens system, I will return to deriving Eq. (9.12).

9.3 – Optical Power of a two-lens system

To determine the power of a two-lens system consider Fig. 9.16, We will analyze it as if the lenses were thin.

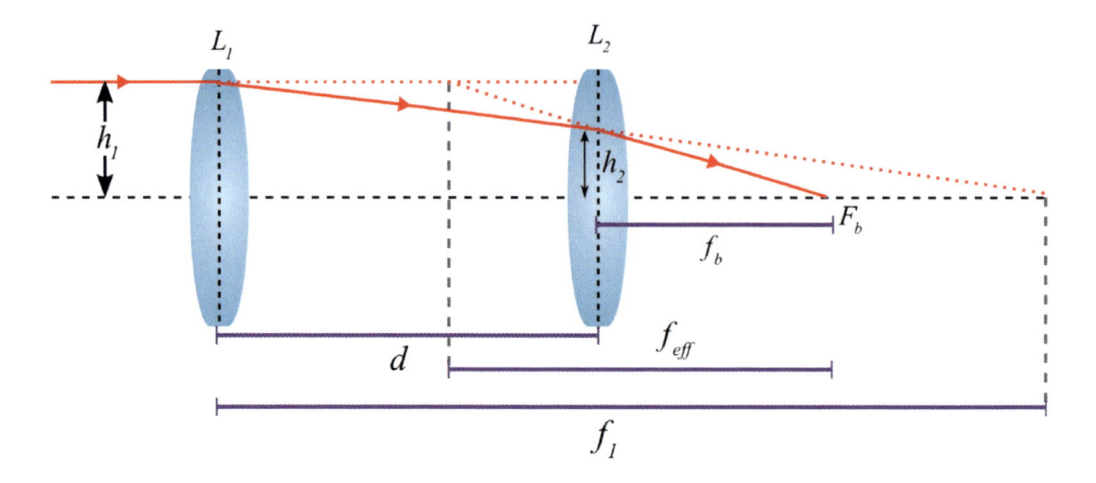

Figure 9.16 – A two-lens system, with thin lens parameters.

We have one lens with focal length f_1 followed by another with focal length f_2 and they are a distance d apart. Light arriving at the first lens from far away hits it at height h_1 with the intention of focusing a distance f_1 to the right, however it is intercepted by the second lens at a height h_2, and is refracted once again finally crossing the symmetry axis a distance f_b to the right of the second lens. By connecting up the line of the last ray with that of the incoming ray we identify the effective plane at which the transition takes place, as if the two lenses were acting as one. Light effectively bends at this principle plane. The distance from this plane to the overall point of focus is the focal length f_{2tnl} of the two-lens system. Our interest is in relating f_{2tnl} to f_1, f_2 and d. To do this h_1 and h_2 will need to be eliminated. It is convenient to first determine f_b. The point that the initial ray was heading toward after being refracted by the first lens will become a virtual object for the second lens. From Eq. (9.9) we have

$$\frac{1}{f_2} = \frac{1}{-(f_1-d)} + \frac{1}{f_b},\qquad(9.13)$$

from which the back focal length f_b is found;

$$\frac{1}{f_b} = \frac{1}{f_2} + \frac{1}{(f_1-d)}.\qquad(9.14)$$

Now we must relate f_{2tnl} to f_b. Two pairs of similar triangles aid in this and provide two equations for the ratio h_2/h_1;

$$\frac{h_2}{h_1} = \frac{f_1-d}{f_1} = \frac{f_b}{f_{2tnl}}.\qquad(9.15)$$

Combine Eq. (9.15) with Eq. (9.14) provides an equation for $1/f_{2tnl}$,

$$\frac{1}{f_{2tnl}} = \frac{1}{f_1} + \frac{1}{f_2} - \frac{d}{f_1 f_2}.\qquad(9.16)$$

This equation will have a great deal of utility for a host of two-lens systems. By re-expressing Eq. (9.16) in terms of lens power, we arrive at the realization that when thin lenses are sandwiched close enough together lens powers essentially add. More generally

$$P = P_1 + P_2 - P_1 P_2 d.\qquad(9.17)$$

9.4 – Matrix Optics

When there are many surfaces in an optical problem (multiple lenses, mirrors, etc.), simple ray tracing to follow rays of light becomes an extreme burden. A way around this is to turn linearized paraxial theory into a matrix form. To see how this works we reexamine refraction by a spherical surface, see Fig. 9.17.

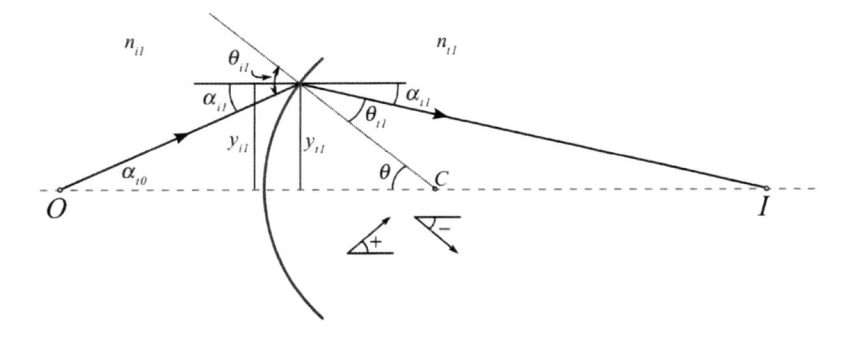

Figure 9.17 – Refraction from a single spherical surface.

Fig. 9.17 shows most of the important parameters. A ray traveling upward will have its direction measured by its angle from the horizontal, α_{t0}. For example the ray incident on the spherical surface in the figure is at an angle α_{i1}, which is considered positive. After refraction the ray heads downward at an angle, α_{t1}, which is negative. Now it is just a matter of rewriting Snell's equation for the paraxial case. For this purpose $\sin(\theta_{i1})$ and $\sin(\theta_{t1})$ in Snell's equation are simply turned into angles in radians, and it is recognized that the height as which the ray hits is the same on either side of the boundary

$$n_{i1}\theta_{i1} = n_{t1}\theta_{t1} \text{ and} \tag{9.18}$$

$$y_{i1} = y_{t1}. \tag{9.19}$$

Translating between symbols gives

$$\theta + \alpha_{i1} = \theta_{i1} \text{ and} \tag{9.20}$$

$$\theta + \alpha_{t1} = \theta_{t1}. \tag{9.21}$$

On this basis

$$n_{i1}(\theta + \alpha_{i1}) = n_{t1}(\theta + \alpha_{t1}) \text{ or} \tag{9.22}$$

$$n_{t1}\alpha_{t1} = n_{i1}\alpha_{i1} - (n_{t1} - n_{i1})\theta. \tag{9.23}$$

By treating θ as a small angle we can write it as

$$\theta \approx y_{i1}/R_1. \tag{9.24}$$

So by putting Eq. (9.23) and Eq. (9.24) together we get

$$n_{t1}\alpha_{t1} = n_{i1}\alpha_{i1} - (n_{t1} - n_{i1})y_{i1}/R_1. \tag{9.25}$$

If we now define the refracting power of the surface to be

$$D_1 = (n_{t1} - n_{i1})/R_1, \tag{9.26}$$

then Eq. (9.18), Eq. (9.19), and Eq. (9.25) describe how the ray changes at the interface,

$$n_{t1}\alpha_{t1} = n_{i1}\alpha_{i1} - D_1 y_{i1} \text{ and} \tag{9.27}$$

$$y_{i1} = y_{t1}. \tag{9.28}$$

These simple linear equations are easiest if written as a (2×2) matrix operating on a (2×1) vector;

$$\begin{bmatrix} n_{t1}\alpha_{t1} \\ y_{t1} \end{bmatrix} = \begin{bmatrix} 1 & -D_1 \\ 0 & 1 \end{bmatrix} \begin{bmatrix} n_{i1}\alpha_{i1} \\ y_{i1} \end{bmatrix}. \tag{9.29}$$

The (2×2) matrix is known as a refraction matrix R_1. Beyond this we need a way to translate a vector from one height to another in a homogeneous medium. Fortunately the ray moves along a straight line in a given medium. Suppose the ray depicted as originating from the axis in Fig. 9.17 actually started from an object y_o high. Then in going from y_o to y_{i1} over a longitudinal distance d_{1o},

$$y_{i1} = y_o + \alpha_{i1} d_{1o}. \tag{9.30}$$

Another (2×2) matrix used to describe lenses is known as a translation matrix T_{1o},

$$\begin{bmatrix} n_{i1}\alpha_{i1} \\ y_{i1} \end{bmatrix} = \begin{bmatrix} 1 & 0 \\ \dfrac{d_{1o}}{n_o} & 1 \end{bmatrix} \begin{bmatrix} n_o \alpha_o \\ y_o \end{bmatrix}, \tag{9.31}$$

that moves the ray from the object o to surface 1 (Note that the subscript on the matrix is written from right to left). By this scheme of matrices a complete lens starts with an incident vector at surface 1 followed by translation from surface 1 to surface 2, and that followed by refraction at surface 2. Written from right to left, the lens is represented as

$$\begin{bmatrix} n_{t2}\alpha_{t2} \\ y_{t2} \end{bmatrix} = \begin{bmatrix} 1 & -D_2 \\ 0 & 1 \end{bmatrix} \begin{bmatrix} 1 & 0 \\ \dfrac{d_{21}}{n_{t1}} & 1 \end{bmatrix} \begin{bmatrix} 1 & -D_1 \\ 0 & 1 \end{bmatrix} \begin{bmatrix} n_{i1}\alpha_{i1} \\ y_{i1} \end{bmatrix}. \tag{9.32}$$

The product of the three (2×2) matrices is known as the system matrix of the lens. The system matrix acts on the incident ray at surface 1 and yields the refracted ray (transmitted) at surface 2,

$$\begin{bmatrix} n_{t2}\alpha_{t2} \\ y_{t2} \end{bmatrix} = \begin{bmatrix} 1 - \dfrac{D_2 d_{21}}{n_{t1}} & -D_1 - D_2 + \dfrac{D_1 D_2 d_{21}}{n_{t1}} \\ \dfrac{d_{21}}{n_{t1}} & 1 - \dfrac{D_1 d_{21}}{n_{t1}} \end{bmatrix} \begin{bmatrix} n_{i1}\alpha_{i1} \\ y_{i1} \end{bmatrix}. \tag{9.33}$$

An image of a lens that models with such a matrix operation is shown in Fig. 9.18. The elements of the system matrix contain all of the paraxial information concerning the lens. Identifying them by row and column as

$$\begin{bmatrix} a_{11} & a_{12} \\ a_{21} & a_{22} \end{bmatrix} = \begin{bmatrix} 1-\dfrac{D_2 d_{21}}{n_{t1}} & -D_1 - D_2 + \dfrac{D_1 D_2 d_{21}}{n_{t1}} \\ \dfrac{d_{21}}{n_{t1}} & 1-\dfrac{D_1 d_{21}}{n_{t1}} \end{bmatrix} \qquad (9.34)$$

provides me with less to type.

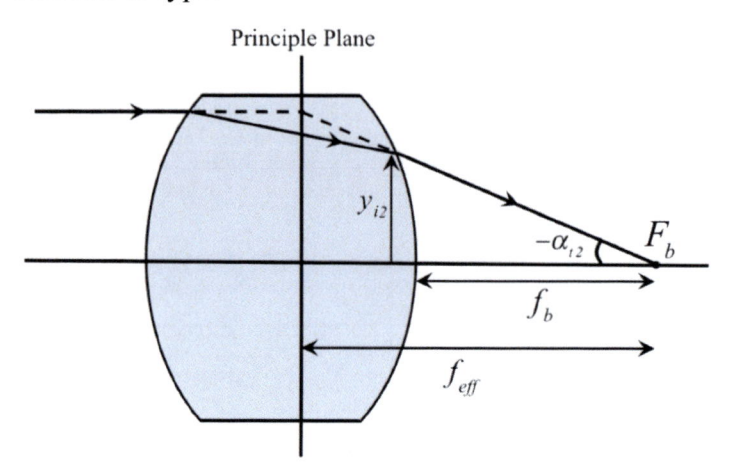

Figure 9.18 – A thick lens diagram.

To start in I will send a ray in from far away ($\alpha_{i1} = 0$). The elements of the transmitted ray are then $n_{t2}\alpha_{t2} = a_{12} y_{i1}$ and $y_{t2} = a_{22} y_{i1}$. A ray heading downward at an angle α_{t2} from a height y_{t2} will reach the symmetry axis in a longitudinal distance $y_{t2}/(-\alpha_{t2}) = a_{22} y_{i1}/(-a_{12} y_{i1}/n_{t2})$. For a lens in air this back focal length is $f_b = a_{22}/(-a_{12})$. If we extend the ray backward from the back focal point then it will go a distance $y_{i1}/(-\alpha_{t2}) = y_{i1}/(-a_{12} y_{i1}) = 1/(-a_{12})$ before meeting the incident ray direction. This distance is the effective focal length, and consequently

$$\frac{1}{f_{eff}} = D_1 + D_2 - \frac{D_1 D_2 d_{21}}{n_{t1}} = (n_l - 1)\left[\frac{1}{R_1} - \frac{1}{R_2} + \frac{(n_l - 1)d_l}{R_1 R_2 n_l}\right], \qquad (9.35)$$

where I have changed the refractive index between the surfaces from n_{t1} to n_l, with the subscript *l* being used for *lens*. You will note that this is the focal length

I promised to derive. On this basis the back focal length and the effective focal length are found to be proportional.

We can easily obtain the system matrix for a thin lens from the thick lens matrix [Eq. (9.34)] by setting $d_{21} = 0$;

$$\begin{bmatrix} 1 & -\frac{1}{f} \\ 0 & 1 \end{bmatrix},\qquad(9.36)$$

where f is the thin lens focal length [Eq. (9.10)].

9.5 – Chapter Nine Exercises

Exercise 9.1 – A light source is placed at one of the focal points of a mirrored ellipse and a detector is place at the other focal point as shown. An absorbing surface (blue) is placed between the source and detector so that light cannot transmit directly between the two.

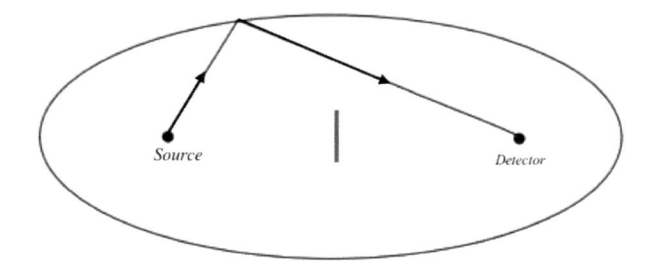

The inner surface of the ellipse is perfectly reflecting. If the semi-major axis is A and the semi-minor axis is B, what is the overall length for which light takes the least time to go from the source to the detector. The index of refraction within the ellipse is 1. Hint: the light must reflect off the mirror (a sample path is shown).

Exercise 9.2 – Show that all our (2×2) matrix optics matrices have determinants of 1.

Exercise 9.3 – *Single Thin Lens* – Find the focal length of a glass lens ($n = 1.5$) with $R_1 = 20\ cm$ and $R_2 = 40\ cm$.

Exercise 9.4 – Find the refractive index that makes a cat's eye retro-reflector work. A cat's eye retroreflector is a spherical lens that is mirrored at the back and reflects light parallel to the light that enters it.

Exercise 9.5 – Trace the cat's eye with matrices to prove that if it has the refractive index in Exercise 9.4 then the light returns parallel to the incident light.

Exercise 9.6 – Find the paraxial focal length of spherical lens of refractive index $n = 1.45$ and diameter 2 cm in air.

Exercise 9.7 – *Thick Lens* – Show that the paraxial focal length of a spherical lens with refractive index n and diameter D in air is

$$f_{eff} = \frac{nD}{4(n-1)}.$$

Hint: See Eq. (9.12).

Exercise 9.8 – *Two Thin Lenses* – Two thin lenses are a place 5 cm from each other. The first lens (lens 1) has focal length $f_1 = 20$ cm and the second (lens 2) has focal length $f_2 = 25$ cm. What is the focal length f of the system?

Exercise 9.9 – Find the matrix for a biconvex lens with index of refraction 1.5, thickness 5 cm, and radius ± 20 cm for the respective sides.

Exercise 9.10 – The system matrix for a thick biconvex lens in air is given by

$$\begin{bmatrix} 0.6 & -0.26\,cm^{-1} \\ 2.0\,cm & 0.8 \end{bmatrix}.$$

Knowing that the first radius is 5 cm, that the thickness is 3 cm, and that the index of refraction of the lens is 1.5, find the second radius.

Exercise 9.11 – Can one make a perfect lens, i.e. a lens that has no spherical aberration? Hint: Try a plano-convex configuration. What shape would the convex surface have to take if the lens has only 1 focal point independent of the height h of the incident rays? A good approach to calculating the shape is Fermat's principle. The calculation is similar to the one in Fig. 9.5, but with refractive index controlling the velocity of light.

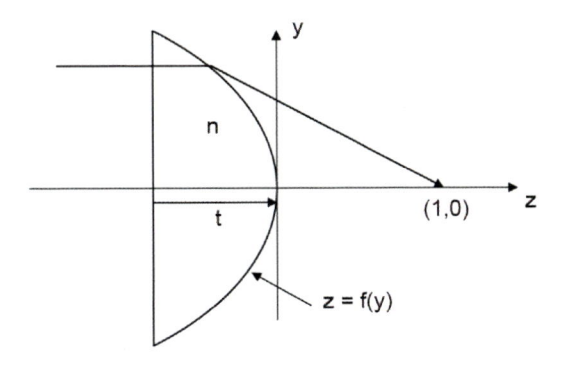

Exercise 9.12 – Calculate the paraxial system matrix elements for the thick spherical lens shown below. It has a radius $R = 1$ mm, a refractive index $n = 1.45$, and is surrounded by air.

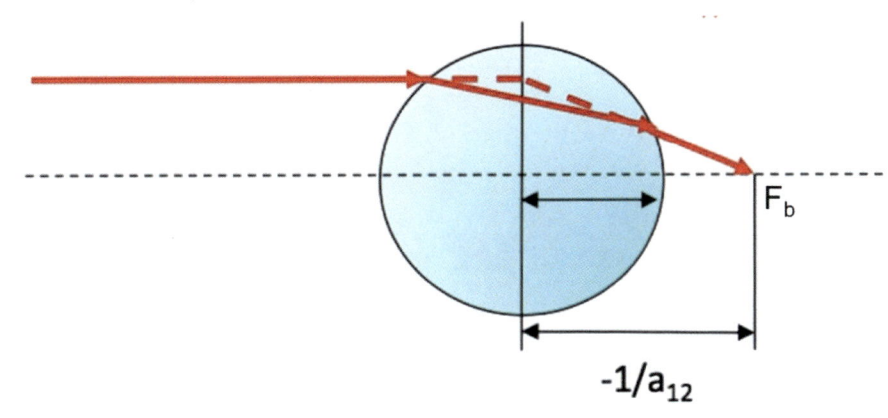

Exercise 9.13 – Apply matrix optics to the cornea of the human eye in Fig. 9.19 in order to determine where it would focus if there were no crystalline lens. The cornea is the front layer about 449 microns thick and having a refractive index of 1.376, with air on the outside (anterior) and an aqueous like medium on its posterior side (i.e. take the refractive index of this medium to be 1.336). With the two spherical interfaces shown and the radii given (R_1, $R_2 = 7.259$mm, 5.585mm) and **without** the crystalline lens, determine where the light from an infinitely distant object focuses, as measured from the front of the cornea. If your answer is greater than the eye's axial lenth of 24 mm, then imaging on the retina is impossible, and the crystalline lens becomes of great importance.

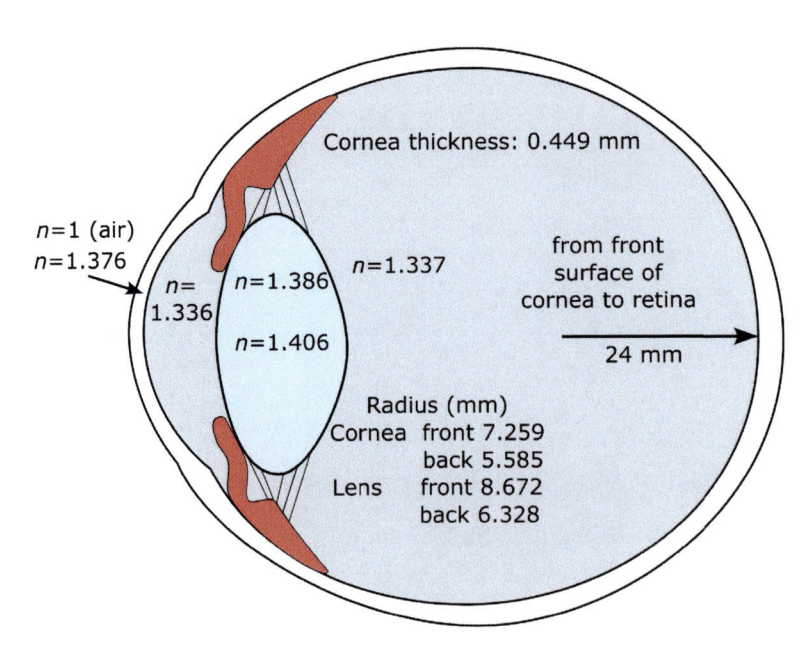

Fig. 9.19 – The physical makeup of a typical human eye.

Experiment #8 – Propagating light through a refractive index gradient

Fill your Pocket Optics container with water. Pour sugar into the vessel without stirring. Allow the sugar to dissolve for several hours (waiting 24 hours or longer may improve results). Make certain that when equilibrium is reached there is still a coating of visible sugar crystals on the bottom. Now send light from your laser horizontally through the container making certain that the incident angle is perpendicular to the container wall. If the sugar is properly dissolved you should see the laser beam bend as shown below. Does the curvature of the beam depend on the beam elevation above the sugar crystals? How does the curvature of the beam vary with the height at which the beam enters the solution? What can you tell from this observation about the distribution of dissolved sugar in the water?

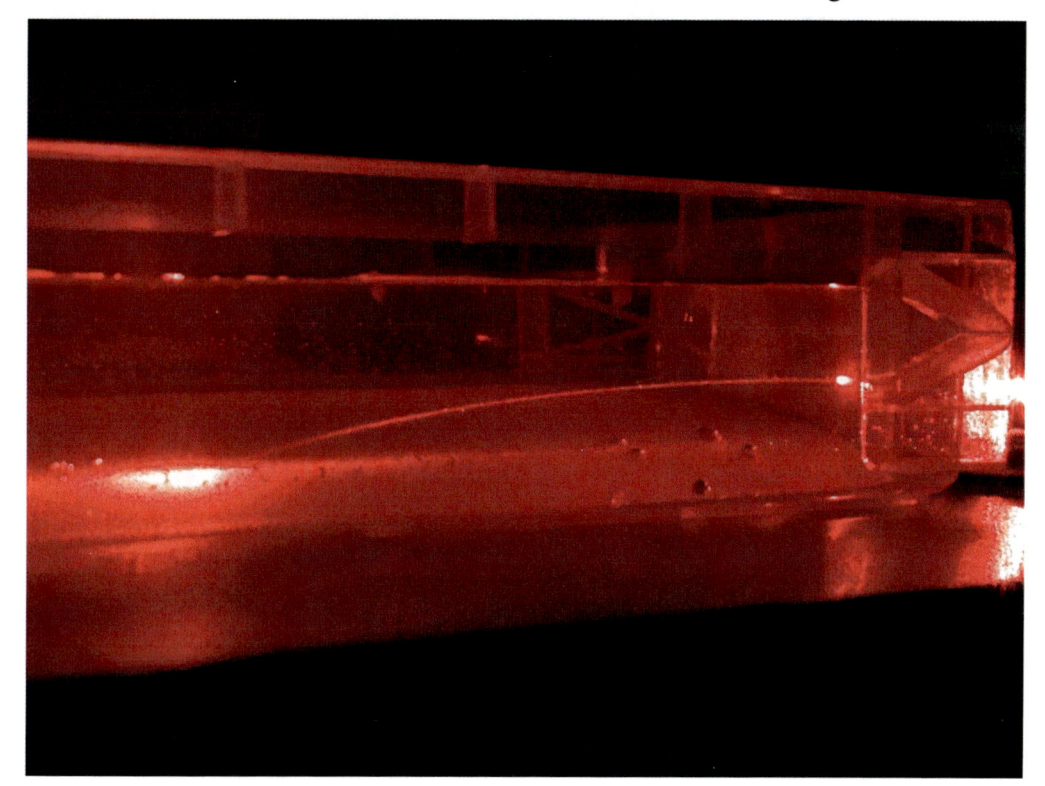

Scan for a video demo of this experiment

Chapter Ten – Optical Instruments

10.1 – The Human Eye

In the first chapter we stated that the range of the electromagnetic spectrum for visible light is from around 400 to 750 nm in wavelength. This is a relatively small portion of the electromagnetic spectrum, which ranges from Extremely Low Frequency Radio waves (ELF) with a wavelength as large as $1 \times 10^8 \, m$ to gamma rays with wavelengths on the order of picometers, $1 \times 10^{-12} \, m$. To get a rough idea of why the visual spectrum may take the range it does lets consider the absorption spectrum of the Earth's atmosphere. Fig. 10.1 shows the spectrum of light absorbed by the atmosphere. It appears that much of the light with energies higher than the visible range (i.e. shorter wavelengths) is absorbed by the Earth's atmosphere.

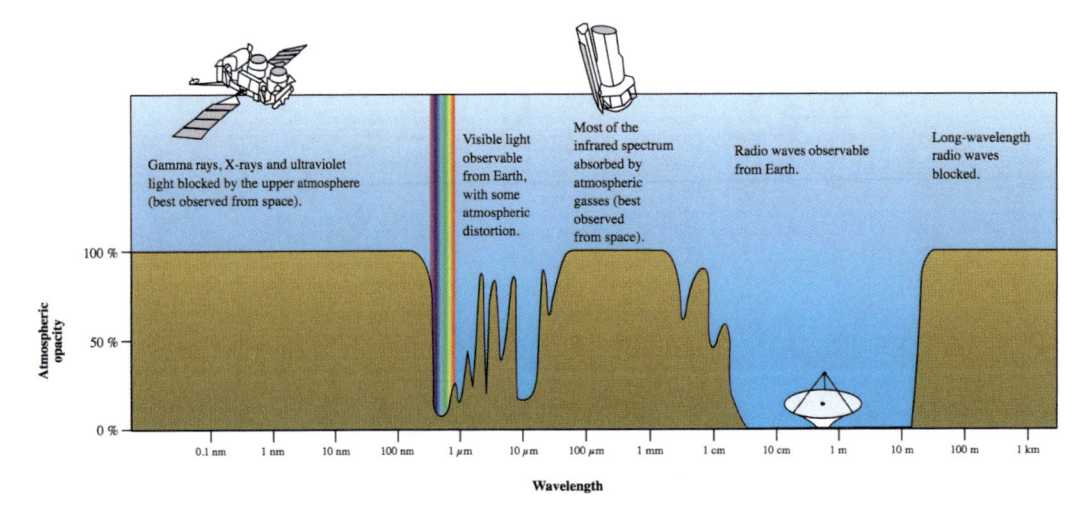

Figure 10.1 – Spectrum of radiation that reaches the Earth's surface (Image credit: NASA).

The diagram of the eye in Fig. 10.2 offers an overview of the main components of the eye. The cornea and the crystalline lens are responsible for focusing light on the retina, which transmits information to the brain via the optical nerve. Unlike the lenses we have previously studied the eye has the remarkable ability to adjust and refocus on objects at different distances from it. The ciliary muscle is responsible for adjusting the crystalline lens to bring objects in and out of focus. The iris is the colored region around the pupil of the eye. The size of the pupil controls the amount of light that enters the eye by opening or closing based on the intensity being viewed. The aqueous humor is a liquid layer between the cornea and crystalline lens that provides protection and supplies nutrients to these areas.

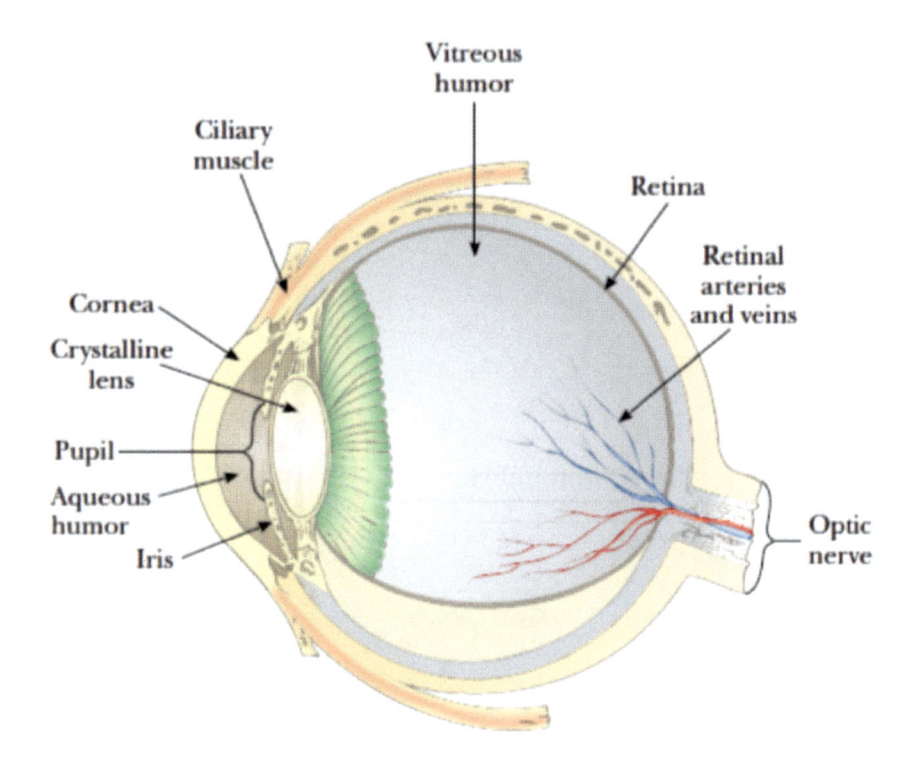

Figure 10.2 – Diagram of the human eye.

The vitreous humor is a transparent, gelatinous substance that helps the eye maintain its shape and keeps the retina in place. The retina is the portion of the eye that distinguishes colors and transports information to the brain as electrical impulses. In fact the retina is directly connected to the brain and some of the information processing occurs before transmission to the optical nerve. The retina is composed primarily[22] of two types of photoreceptor cells called rods and cones. The cone cells are responsible for distinguishing colors and are concentrated near the center of the retina. The rod cells operate in black and white and are active in dim environments.

The cornea supplies most of the eye's optical power, however the crystalline lens adds about 20% to the power of the cornea, and it's power is adjustable. We can find the total optical power of the eye by examining the range of distances at

---

[22] There is an additional photoreceptor called a photosensitive ganglion cell that is responsible for monitoring long-term changes in the external environment such as the length of days.

which focus can be achieved. The closest point at which the eye can be focused is called the near point and the maximum distance at which the eye produces a clear image is called the far point. The near point varies with individuals and changes significantly with one's age. A child can focus on objects as close as 6.5 cm from the eye but as an adult this distance will be much larger. The average human has a near point of 25 cm. Based on a simple model for which both the cornea and crystalline lens are replaced by a single lens of focal length f at the front of the eye, the lens power P is

$$P = \frac{1}{f}. \qquad (10.1)$$

The axial length of the eye is about 2.4 cm If the image is directly on the retina $s_i = 2.4\ cm$ we can find the power of this lens at the near point, $s_o = 25\ cm$, (Fig. 10.3) using a thin lens approximation [derived in Chapter 9]

$$P = \frac{1}{f} = \frac{1}{s_o} + \frac{1}{s_i} = \frac{1}{0.25\ m} + \frac{1}{0.024\ m} \approx 46\ D. \qquad (10.2)$$

For a young person with a near point of 6.5 cm the power is 57 D. If we consider the far point to be much greater than a meter we can ignore the first term in the equation and find

$$P = \frac{1}{s_i} = \frac{1}{0.025\ m} \approx 42\ D. \qquad (10.3)$$

This is the power of a relaxed eye based on our simple model. In this analysis we have treated the cornea and crystalline lens system as a single thin lens. For most purposes this is sufficiently accurate to understand the optical system, it can even model the conditions of nearsightedness, or myopia, and farsightedness, or hyperopia, reasonably well.

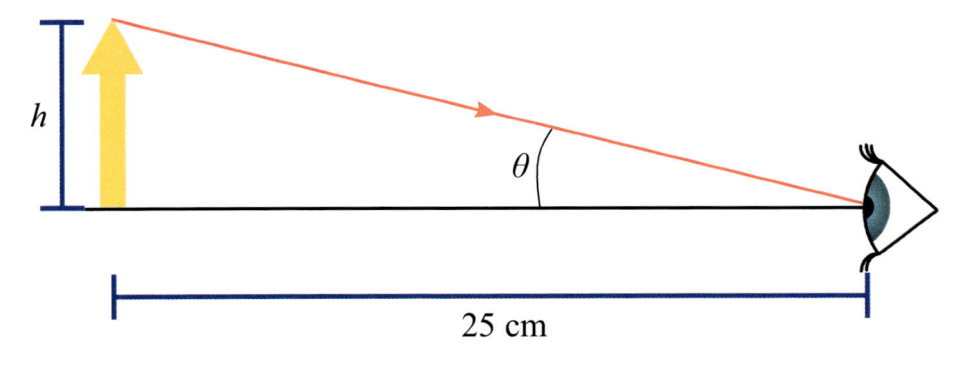

Figure 10.3 – The near point of the average human eye.

Another limitation of the eye is associated with diffraction. Suppose you view two distant points of light such as stars. The stars subtend an angular separation as seen by the eye.

$(\Delta\theta)_{limit} \simeq 2.9 \times 10^{-4}\, rad$

$(\Delta y)_{limit}$

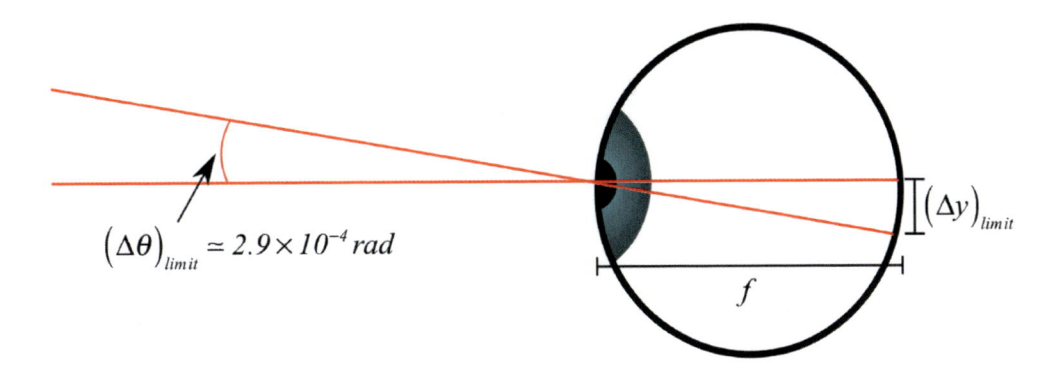

f

Figure 10.4 – Two stars viewed by the eye will not be seen as separate if they subtend an angle between them smaller than $(\Delta\theta)_{limit}$. *Instead of imaging as points on the retina, their images will appear to have a width* $(\Delta y)_{limit}$ *caused by diffraction.*

When this separation is reduced below about 1 minute of arc or 3×10^{-4} radians, separate stars will no longer be seen. To understand this we can return to diffraction by a slit. Although such a pupil shape is more appropriate for a cat than a human, it can give us an estimate. You will recall that the first zero in diffraction intensity from a slit occurs when the sine of the diffracted angle times the slit width is equal to the wavelength. If the two stars are separated by this angle $(\Delta\theta)_{limit}$ or below then they cannot be distinguished. We previously called the slit width a when dealing with diffraction from a slit, so for a small angle of diffraction $(\Delta\theta)_{limit}$, $a(\Delta\theta)_{limit} = \lambda$, and the limiting angle $(\Delta\theta)_{limit} = \lambda/a$. Although a human eye has a circular pupil of diameter D, the equation for the minimum angle is nearly the same

$$(\Delta\theta)_{limit} = \frac{1.22\lambda}{D}. \qquad (10.4)$$

What the diffraction limitation looks like at the retina is shown in Fig. 10.5. For a single star a so-called Airy pattern is seen [Fig. 10.5(a)]. For two stars separated by more than the minimum angle, there are two Airy patterns [Fig. 10.5(b)].

When $(\Delta\theta)_{\lim it}$ is reached the sum of the intensities from each star no longer produce two distinguishable maxima [Fig. 10.5(c)].

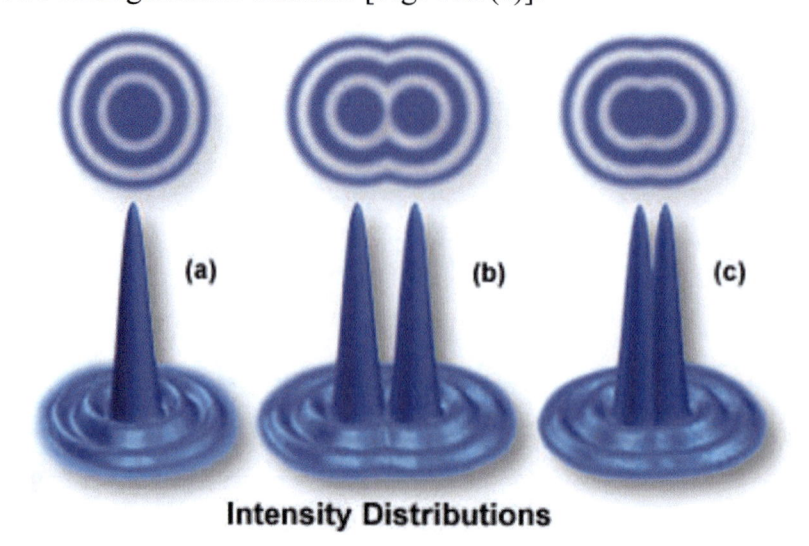

Figure 10.5 – Airy disks.

This separation $(\Delta y)_{\lim it}$ is simply the focal length times $(\Delta\theta)_{\lim it}$

$$(\Delta y)_{\lim it} = \frac{1.22 f \cdot \lambda}{D}. \tag{10.5}$$

As an example: for the human eye let's take the diameter of the pupil to be 2 mm, and the focal length 2.5 cm. For a wavelength of $\lambda = 500$ nm we find

$$(\Delta y)_{\lim it} = \frac{1.22(2.5 \text{ } cm) \cdot (0.5 \times 10^{-4} cm)}{0.2 cm} \approx 7.5 \times 10^{-4} cm = 7.5 \mu m. \tag{10.6}$$

From the geometry of Fig. 10.4 we can find the smallest separation that can be resolved for objects distance x from the eye. The minimum resolvable distance d between two points is given by

$$d = x \tan\left((\Delta\theta)_{\lim it}\right) \approx x(\Delta\theta)_{\lim it}. \tag{10.7}$$

For an object at the near point of the human eye (25 cm) the minimum resolvable separation d is $25 cm \cdot (\Delta\theta)_{\lim it} = 75 \mu m$, 10 times $(\Delta y)_{\lim it}$. This is an upper limit, since the pupil diameter can be larger, and filtered light in the blue makes the wavelength smaller; both changes will increase the resolution (i.e. reduce d).

The retina display on your iPhones is designed with this in mind. The pixels are sized around 80 microns for most Apple touch screen devices; the smallest pixel size is found in the Retina HD display of the iPhones 6, 6S, 7 and is 63 microns. The engineers at Apple realized that pixels smaller than this would not increase the apparent image quality because the human eye would be unable to resolve them at the distances at which we normally hold our phones. However, the pixels are readily viewed using a magnifying lens attached to the camera lens of another phone. By attaching a lens to my phone as described in Experiment 9, I was able to obtain this photo of the LCD display of another iPhone 5S, Fig. 10.6.

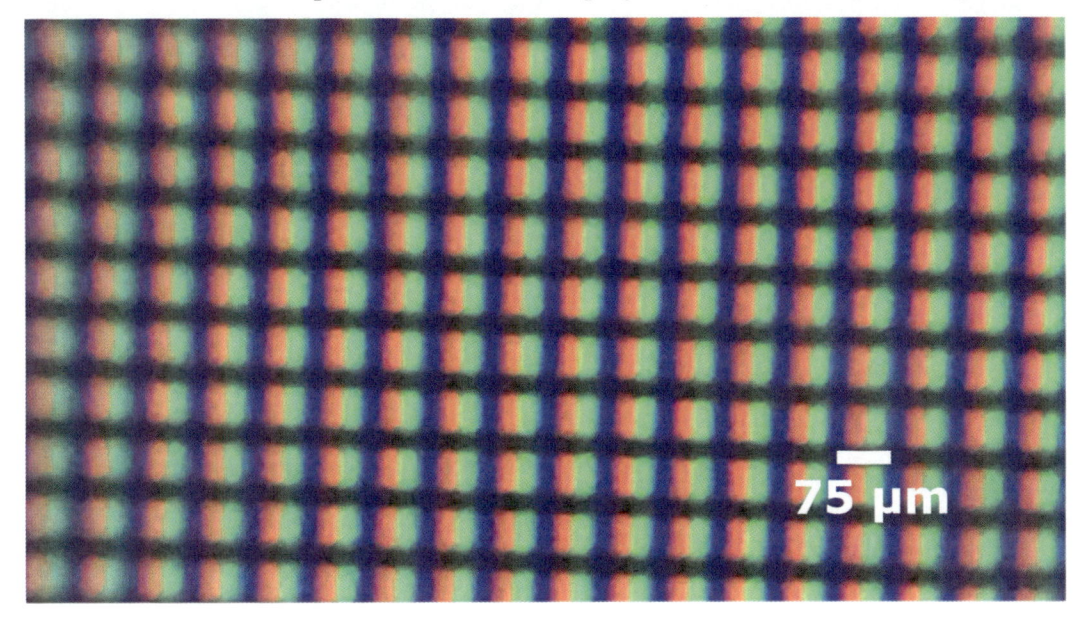

Figure 10.6 – An image of an iPhone 5S screen taken using a magnifying lens on another iPhone.

10.2 – The Magnifying Glass

The minimum resolvable separation (MRS) can be reduced below the naked eye resolution of ~ 75 μm with the aid of optical instruments. In order to view a small object in greater detail one may employ a magnifying glass. Let's first discuss the magnifying glass and afterwards return to glasses and contacts.

Optical lenses came into use quite early in human history. They are mentioned in a 5th-century BC, Greek play, "The Clouds", written by Aristophanes, as a tool purchased from a druggist to kindle fires. The lenses were employed for medical purposes and used to cauterize wounds and their ability to magnify images was noted. However, it was not until the 13th-century that Roger Bacon described the mechanism. Bacon built on the theory of optics that had been developed by the Arab scholar Ibn al-Haytham in his "Book of Optics". The theories of reflection and refraction had been described by al-Haytham, and Bacon recognized that the bending of light by a lens was a result of these phenomena and developed the magnifying glass for scientific purposes.

The magnifying glass is used to see details that the unaided eye cannot resolve at its near point ($s_o \sim 25cm$). An object of height h subtends an angle θ_0 as shown in Fig. 10.7. For small angles, and for an object at the near point,

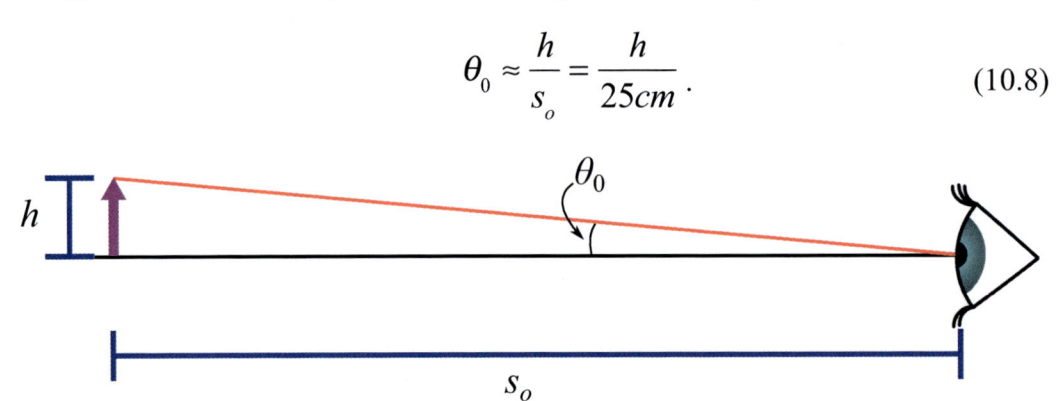

$$\theta_0 \approx \frac{h}{s_o} = \frac{h}{25cm}. \qquad (10.8)$$

Figure 10.7 –The angle θ_0 subtended by an object of height as seen by a naked eye. .

A single lens can help one resolve an object that is too small to view with the naked eye. Placing an object inside the focal point f of a converging lens (Fig. 10.8) produces a virtual image of the original object that subtends an angle θ larger than the angle formed by the naked eye (θ_0). The angular magnification provided by the lens is

$$m_\theta = \frac{\theta}{\theta_0}. \qquad (10.9)$$

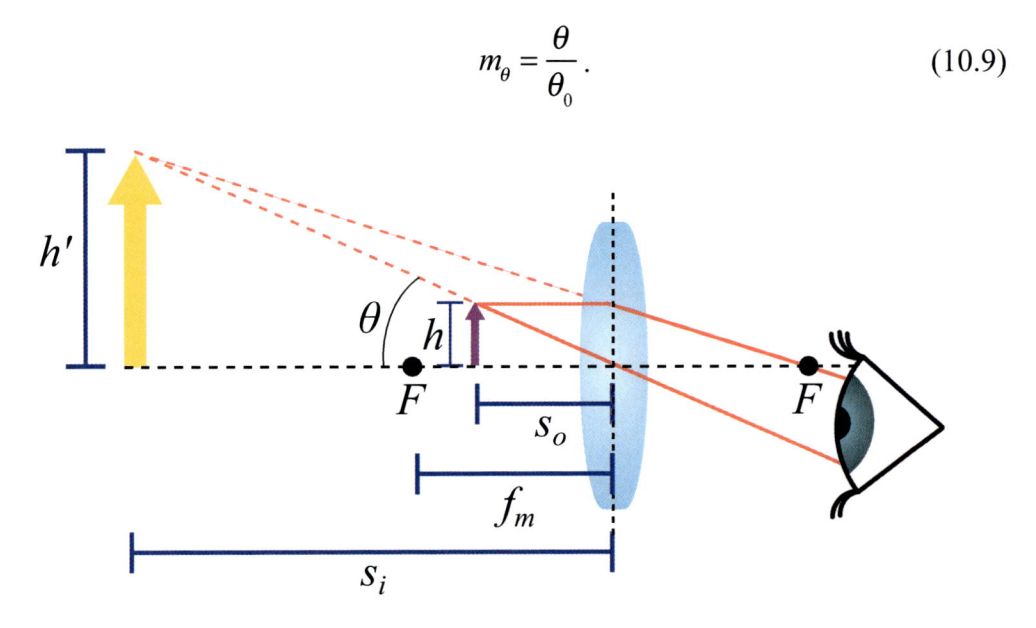

Figure 10.8 – A magnifying lens, ray traced as thin, creates an image of an object and can increase resolution of the object.

The object distance s_o in Fig. 10.8 can be found using the thin lens equation for the magnifier, with the image distance $s_i = -25\ cm$ (negative because it is virtual) we have

$$\frac{1}{s_o} + \frac{1}{s_i} = \frac{1}{s_o} - \frac{1}{25\ cm} = \frac{1}{f_m}; \quad s_o \simeq \frac{h}{\theta}; \quad \frac{\theta}{h} \simeq \frac{1}{f_m} + \frac{1}{25\ cm}. \quad (10.10)$$

So the new viewing angle is

$$\theta \simeq h\left[\frac{1}{f_m} + \frac{1}{25\ cm}\right], \quad (10.11)$$

which is considerably larger than the angle for the unaided eye since f_m is considerably smaller than $0.25\ m$. This means a smaller h can be resolved more easily with the magnifier than with the eye alone. The angular magnification provided by the lens is

$$m_\theta = \frac{\theta}{\theta_0} = \frac{h\left[\frac{1}{f_m} + \frac{1}{25\ cm}\right]}{\frac{h}{25\ cm}} = \frac{25\ cm}{f_m} + 1. \quad (10.12)$$

A reasonable magnification for a simple "Drug Store Magnifier" is ~ 4. Beyond this angular magnification distortion is increased. Since the resolution of the eye remains at ~1 minute of arc (Fig. 10.4), the minimum resolvable separation (MRS) with a simple magnifier reduces to $75\ \mu m/4 \simeq 19\ \mu m$. As in the case of the naked eye resolution this is an upper limit associated with a simple magnifier (i.e. $m_\theta = 4$).

This is by no means the limit of what is possible by using a single magnifier. Antonie van Leeuwenhoek (1632-1723) achieved an MRS near 1 micron by using a small lenses with very short focal lengths ~1mm, and with angular magnification as large as $266\times$ (J. van Zuylen, 1981). Leeuwenhoek is thought to be the first to employ his "simple" microscope for the study of microorganisms. He is credited with discovering bacteria, the vacuole of the cell, and spermatozoa, among other things. Since no photo could be taken at the time, the historical record does not reveal the image distortion by his system.

An easy way to go beyond this is to use two lenses, the so-called compound microscope.

10.3 – The Compound Microscope

The simple compound microscope is composed of two lenses and allows viewing of objects that are far too small to be seen with the naked eye or a "Drug Store Magnifier". The invention of the compound microscope is a somewhat disputed subject. The discovery is commonly attributed to the Zacharias Janssen, a Dutch spectacle maker, who is claimed to have discovered it in 1590. Janssen combined two lenses, the lens near the object, called the objective lens, created a real image that was further magnified by the second lens, or eyepiece (Fig.10.9). The eyepiece acts as the magnifier we had discussed in the last section. Leeuwenhoek's contemporary, Robert Hooke, used the compound microscope to study biological samples, notably fleas and cork. Hooke discovered plant cells while examining a sample of cork. It should be noted that Leeuwenhoek's "simple microscope" based on a single short focal length magnifier lens is attributed with producing clearer images with higher magnification than Hooke's early compound microscope. It was not until the mid-1800s that further improvements were made to the compound microscope. A German engineer, Carl Zeiss, founded a business[23] that sold and improved optical instruments. Zeiss employed scientists and apprentices to produce higher quality lenses. A physicist, Ernst Abbe was appointed research director for the business. Abbe invented many optical instruments and developed equations to describe the working of optical instruments. Abbe was the first person to define the numerical aperture (NA) of an optical device, which we will discuss in the next section. He also discovered a physical limit of resolution that can be achieved using lenses.

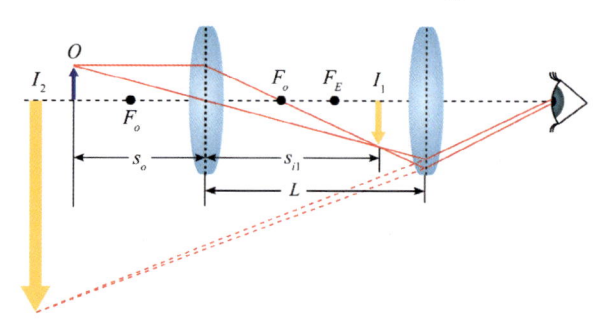

Figure 10.9 – The two-lens system of a compound microscope, ray traced as if thin.

---

[23] The ZEISS company is still actively creating and selling optical instruments today.

The two-lens system of the microscope is shown in Fig. 10.9. The first lens creates a real image, I_1, of the object, O. The second lens then magnifies I_1 and the observer sees the virtual image I_2. The system can be analyzed by treating the image of the object created by the first lens as an object being imaged by the second lens. The magnification provide by the system is the product of the magnifications of the objective and eyepiece,

$$M = M_o m_\theta = \frac{s_{i1}}{s_o}\left(\frac{25\,cm}{f_e} + 1\right) \approx -\frac{L}{f_o}\left(\frac{25\,cm}{f_e}\right), \qquad (10.13)$$

where M_o is the magnification of the objective and the final result is approximate (i.e. it assumes that the real image formed near the end of the microscope barrel length, L). The resolution, or smallest separation for which objects can be separately discerned, is limited by diffraction (see next section). The smallest separation that can be seen in blue light with a compound microscope is $\sim 0.2\ \mu m$, about one hundred times smaller than what can be seen with a simple magnifier.

10.4 – Abbe's Diffraction Limit

In tuning and improving optical instruments such as microscopes it seems natural that improvements in technology will allow for imaging of arbitrarily small objects using light that is reflected from or transmitted by a sample. However, this is not the case and towards the end of the 19th century Ernst Abbe, an expert on optical instruments, recognized a fundamental limit on the smallest separation between two objects that can be resolved by a lens. Deriving this limit involves a complicated analysis of the diffraction that occurs when light travels through a lens aperture and we will not go through that derivation here. The result Abbe achieved is associated with the numerical aperture (NA) of the lens a quantity Abbe defined to be

$$NA = n\sin(\alpha), \qquad (10.14)$$

where n is the refractive index of the external medium and α is the angle formed in Fig. 10.10.

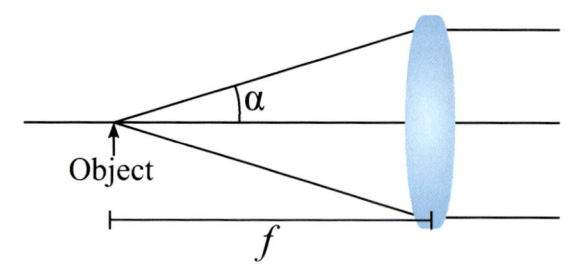

Figure 10.10 – Lens focused on an object.

The smallest separation d that can be imaged according to the diffraction limit is

$$d = \frac{\lambda}{2(NA)} = \frac{\lambda}{2n\sin(\alpha)}. \qquad (10.15)$$

A very good lens may have $NA \approx 1$, thus the smallest object that can be imaged is approximately half the wavelength of the light used to create the image. The smallest wavelength in the visible spectrum is blue light with $\lambda \approx 400\,nm$ and according to Eq. (10.15) the smallest separation that can be imaged is then $d \approx 200\,nm$.

Advances beyond Abbe's resolution limit were not made for more than a century. However, in the 20th century several improvements were made in biological imaging, namely total internal reflection fluorescent microscopy and confocal microscopy.

10.5 – Total Internal Reflection Fluorescence Microscopy (TIRFM)

We discussed the physics of total internal reflection in Chapter 5 and here we discuss its application to microscopy. In typical fluorescence devices a laser or similarly bright source is used to stimulate a molecule to fluoresce. One of the problems with these devices is that the molecules in the sample do not absorb most of the incident light. The unabsorbed light is detected through reflection or transmission, and produces an unwanted background that the fluorescence has to compete against. A solution to this problem can be devised using the evanescent field created by total internal reflection, as we discussed in Chapter 5.

Fig. 10.11 is a diagram of such a device. Blue light incident on a prism's internal face at an angle θ beyond the critical angle illuminates a cell having fluorescent dye staining its interior. The evanescent field on the cell side excites the dye molecules, sending broadband green fluorescent photons to the objective lens. In this way the microscope looking from above sees very little blue light.

Here we will simply state the lateral resolution that can be achieved by TIRFM, which is

$$d = \frac{\lambda}{2NA}.\qquad(10.16)$$

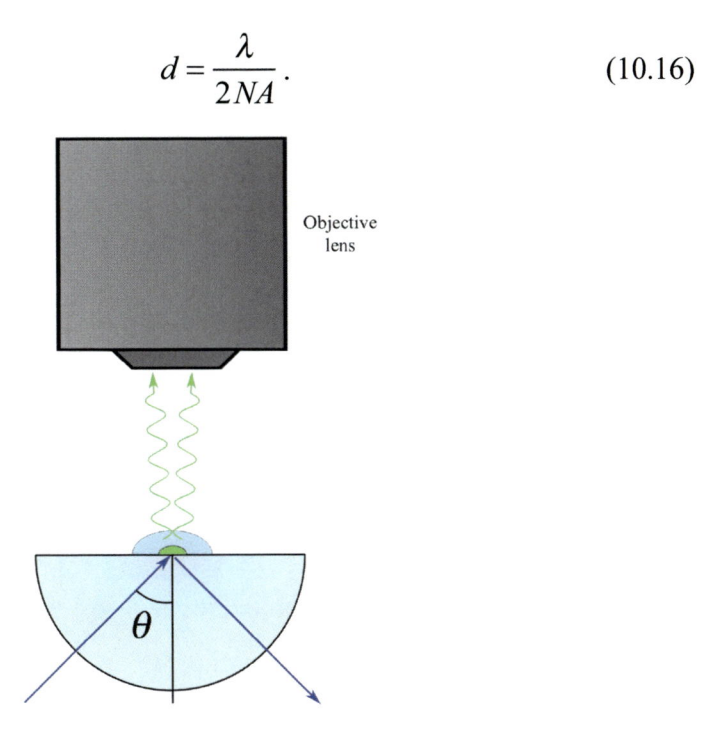

Figure 10.11 – Total Internal Reflection Fluorescence Stage.

The resolution is limited by diffraction. Additionally, depending on the angle of incidence, the evanescent field can excite only a thin layer, such as the cell wall, which allows processes just in that region to be observed. As the angle is moved closer to the critical angle more of the cell is illuminated. Evanescent illumination has the advantage of discriminating depth, i.e. increasing resolution along the objective's axis. However, the lateral resolution is the same as that of a compound microscope.

10.6 – The Confocal Microscope

In the early 20th-century it appeared that the maximum resolution of optical instruments had been achieved, however various improvements were made in imaging technology for biological specimens and 3-D images, confocal microscopy is one such innovation. Confocal microscopy increases the clarity with depth of a fluorescence microscope by allowing 3-dimensional objects to be imaged in layers of different depths. Traditional microscopes see several layers of an object at once, which decreases clarity. Confocal microscopy allows images from different layers to be isolated and then recombined to form a much clearer final image. It combines many mechanisms that may be unfamiliar so we will first analyze these separately and then describe how they interact as a system. Let's begin by discussing fluorescence.

Fluorescence is light emission that occurs after raising a molecule into an excited state. Fluorescent microscopes image fluorescent light. Fluorescent molecules or fluorophores are usually hydrocarbons or heterocyclic compounds. Fluorophores are introduced into biological samples as fluorescent dyes and selectively bind. This allows regions of a cell or other sample to be imaged selectively. Once the dye has been introduced the fluorophores are raised to an excited state by photons from a laser. The photons used to excite the fluorophores are usually in the ultraviolet or blue range and have energy hf_{ex}. The fluorophore is composed of many atoms and the first excited state has many vibrational modes.[24] The photon raises the molecule's energy to one of the higher vibrational modes of the first excited state and the molecule begins to relax to the lowest vibrational mode of this state. This process occurs over a time period on order of a tenth of a picosecond (10^{-13} s). From this lowest vibrational level of the excited state the molecule will emit a photon of energy hf_{em}. This returns the fluorophore to a vibrational level the ground state energy. Fig. 10.12 shows the excitation and emission process and is called a Jablonski diagram. The levels drawn above each electronic state represent vibrational states of that electronic state.

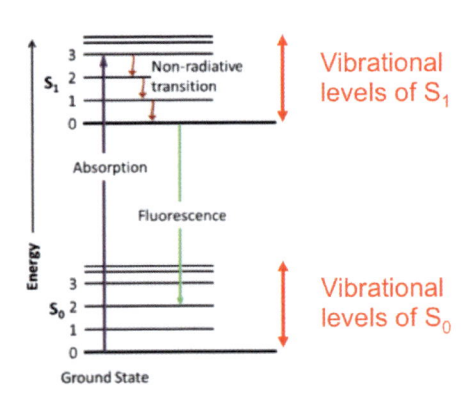

Figure 10.12 – A Jablonski diagram. This shows possible transitions in energy levels or a Fluorophore. (Image credit: The Nobel Foundation)

---

[24] These are related to the number of atoms in the structure.

Notice that the molecule gave some of its energy to its environment before emitting, and therefore the energy of the emitted photon is less than that of the photon used to excite the fluorophore. This difference is attributable to energy that goes into the vibrational modes and corresponds to a difference in the wavelengths of the absorbed and emitted light. An optical filter isolates the light emitted from the fluorophores from the light emitted from the source. Once the excited electron has returned to the ground state energy the fluorophore can absorb another photon. Eventually photolysis occurs destroying the molecular conjugation and permanently bleaching the fluorophore. Because the fluorescence process is cyclical a single fluorophore may emit thousands of photons during the formation of an image. This creates a highly detailed image and ensures that very small regions of the sample can be imaged.

Fluorescence enhances the imaging of biological samples to a considerable degree, however it does not allow us to see the sample in layers as we advertised when introducing the confocal microscope. Viewing the sample in layers requires an additional innovation called a confocal pinhole. The role of the pinhole is to isolate the light from the focal point of the microscope by blocking out adjacent, out-of-focus light. Fig. 10.13 shows how light from the focal region of the sample travels through the pinhole to a second lens.

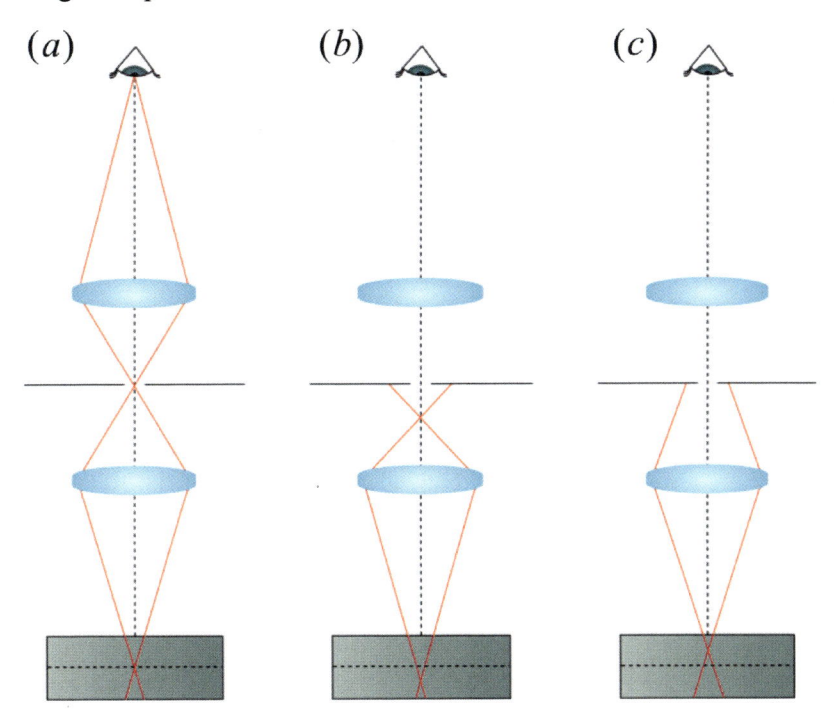

Figure 10.13 – Pinhole used in confocal microscopy.

The pinhole filters out the light from layers above or below the dashed line. Eliminating light from other layers of the sample produces a much clearer image of individual regions of the sample. The device scans over the sample serially producing images of various layers of the sample. The images of the different optical sections are layered to form a 3-dimensional image.

Marvin Minsky of the MIT Media Lab patented the design for the confocal microscope in 1961. An image from the original patent is shown in Fig. 10.14. The critical innovation with this device was not just the use of the confocal pinhole but the scanning device that allowed the image to be constructed serially rather than in parallel (i.e. one point at a time, followed by another, etc.). These points are reconstructed into a single larger image. The mechanism used is called raster scanning and is a process of scanning the sample line-by-line in a pattern. The pinhole design greatly improved the axial resolution of the microscope and allowed for clearer images of 3D biological organisms. However, the lateral spatial resolution for the confocal microscope was still limited to the resolution defined by Abbe, ~ 200 nm.

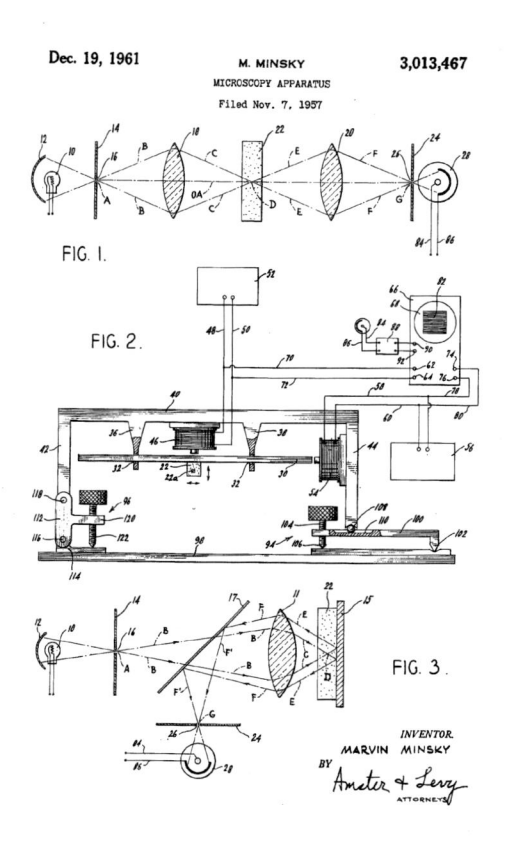

Figure 10.14 – Figures from Patent No. 3013467 filed by Marvin Minsky for the confocal microscope.

10.7 – Breaking the Limit with STED Microscopy

In 1994 a radical proposal was made to overcome the diffraction resolution limit in a fluorescent microscope. This idea, pioneered by Stefan Hell, allows imaging below Abbe's diffraction limit, and is the result of using the laws of quantum physics that had their origin in a 1916 paper published by Einstein[25]. This technology is called stimulated emission depletion (STED) microscopy and relies on controlling the dynamics of the emitting molecules. In discussing STED we will refer back to Chapter 7 for our discussion of stimulated emission and to the previous section for information about fluorescence and confocal microscopy. The STED microscope combines many of the mechanisms we have previously discussed and a diagram of the microscope is shown in Fig. 10.15. The sample in the image is treated with fluorescent molecules and the green (ON) laser is used to excite the molecules in the sample and a much more intense red (OFF) laser is used to stimulate emission from some of the excited fluorophores. The molecules that are not forced to emit by the STED red laser relax naturally, via spontaneous emission, and give off a broad spectrum of light. This light is measured at the detector while the red stimulated emission and green incident light are both filtered out. The size of the region emitting the broadband light, and used to construct the image, is substantially smaller than the Abbe diffraction limit.

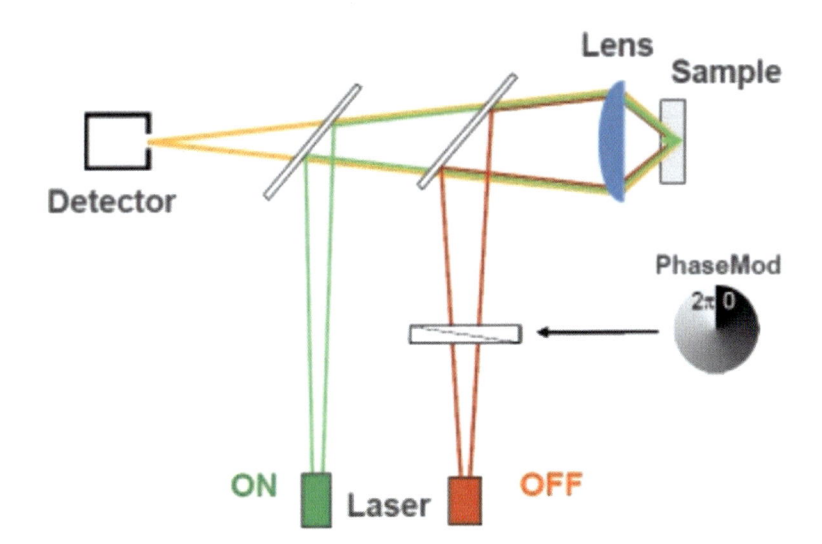

Figure 10.15 - Schematic for the STED microscope. (Image credit: The Nobel Foundation)

---

[25] The 1916 publication "Emission and Absorption of Radiation in Quantum Theory" was expanded in the 1917 paper "Quantum Theory of Radiation".

The improvement in resolution from the STED microscope arises by making the region emitting fluorescence smaller than the region excited by the focused green light, (Abbe diffraction limit). This is done by concentrically exciting the fluorophores with two beams. A green laser that generates an excited region limited by diffraction theory (~200 nm in diameter) and a much more intense red donut shaped beam that spatially eliminates spontaneous emission from the donut region by stimulating excited electrons to the ground state. The excited electrons that are not depleted by stimulated emission emit spontaneously from a much small region than is defined by the diffraction limit, thereby producing a sub-diffraction-limited image. To get an idea of the region mentioned refer to Fig. 10.16.

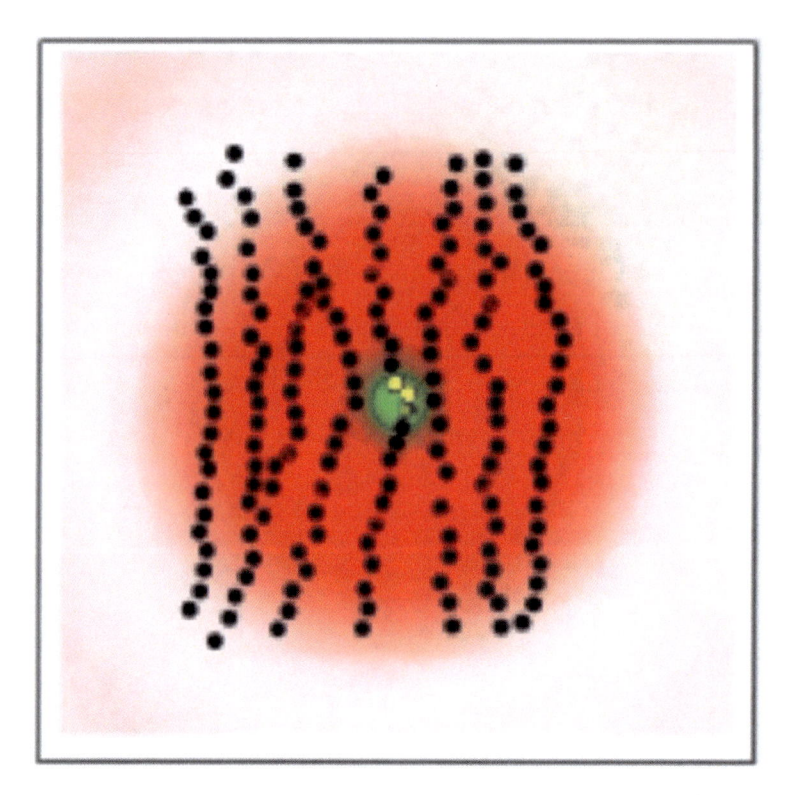

Figure 10.16 – The red region is stimulated to emit using a laser while the green region emits naturally. Without the red laser the region that would emit green would be as large as the red region, which is shown. With stimulation by the red "donut" beam the majority of green emissions within the high intensity portions of the donut are turned off, resulting in a much smaller green emitting diameter. This allows the light from the green region to be isolated and imaged even though it is smaller than Abbe's diffraction limit. (Image credit: The Nobel Foundation)

The difference in the wavelengths of the regions is best understood by examining the energy levels of the fluorophore being used. The energy levels and associated vibrational modes are shown in Fig. 10.17.

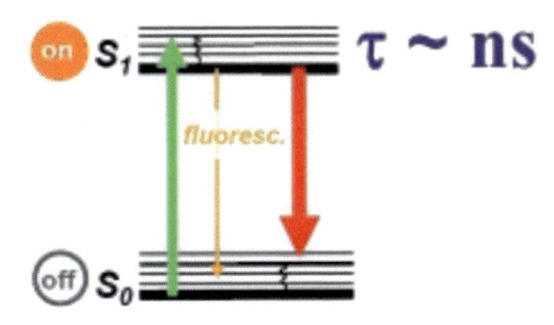

Figure 10.17 – The energy levels and vibrational modes of a fluorophore. (Image credit: The Nobel Foundation)

As we mentioned in the previous section the fluorophore has several electronic energy levels and these energy levels are "clothed" with vibrational modes. Since spontaneous emission is between the lowest vibrational level in the excited state and any of the vibrational levels of the ground state, the emission is generally broad. Stimulate emission on the other hand only occurs at the wavelength of the light source doing the stimulation, and is by comparison monochromatic.

The last important point is that the quality of the image we are able to obtain depends not only on the area of the sample which emits spontaneously but also the intensity of the red laser that is used to stimulate emission in that region. If the intensity is not sufficiently high only a fraction of the molecules in the red region of Fig. 10.16 will be turned off. The fluorophores that remain excited after the red laser pulse will fluoresce naturally and the light they give off will reach the detector and combine with light from the center region, which will decrease the resolution. Fig. 10.18 shows the fluorescent ability of the fluorophores as a function of the intensity of the off beam.

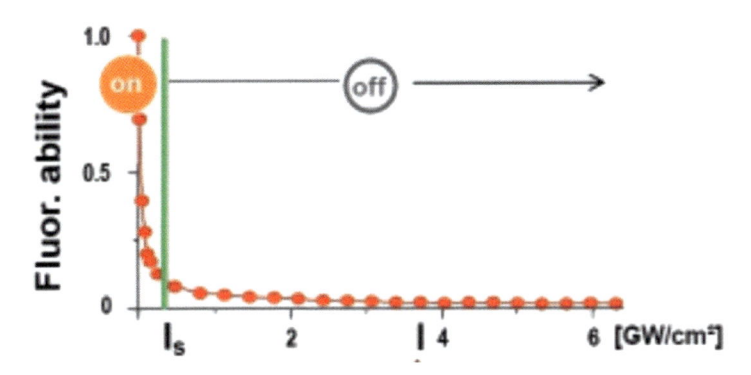

Figure 10.18 – The proportion of molecules in the red region that will fluoresce as a function of the intensity of the off beam. (Image credit: The Nobel Foundation)

When the $I > I_s$ most of the fluorophores in the red region have been emitted via stimulated emission and the region is said to be "off". The new resolution limit has an additional term related to the intensity of the red (OFF) laser I. When the laser exceeds threshold intensity I_s the depletion begins. The STED resolution limit is.

$$d \approx \frac{\lambda}{2n\sin(\alpha)\sqrt{1+I/I_s}}. \qquad (10.17)$$

It is clear from Fig. 10.18 that as I becomes very large only a small fraction of molecules in the red region of Fig. 10.15 will remain in the excited state and fluoresce spontaneously. It is also clear from Eq. (10.17) that as I becomes larger than I_s the minimum resolvable separation d becomes smaller. STED has been used to image separations ~ 0.01 μm, about one twentieth of Abbe's original limit. Next we apply STED to a biological target smaller than 150 nm in size.

A particularly small feature in a cell is the nuclear pore complex. It provides a conduit between the cell's cytoplasm and its nucleus, and many of these complexes are built into the nuclear membrane. The overall size of the nuclear pore complex is 120 nm, but it has a detailed structure. This structure cannot be seen by a confocal microscope, as revealed by the image marked confocal in Fig. 10.19. What is seen is a blur of fluorescent emissions from various dyes that stain the protein that make up the structure. Seen to the right of this blurred image is a STED image of several nuclear pore complexes. Here the resolution difference is apparent, with all 8 of the proteins surrounding a given pore being visible.

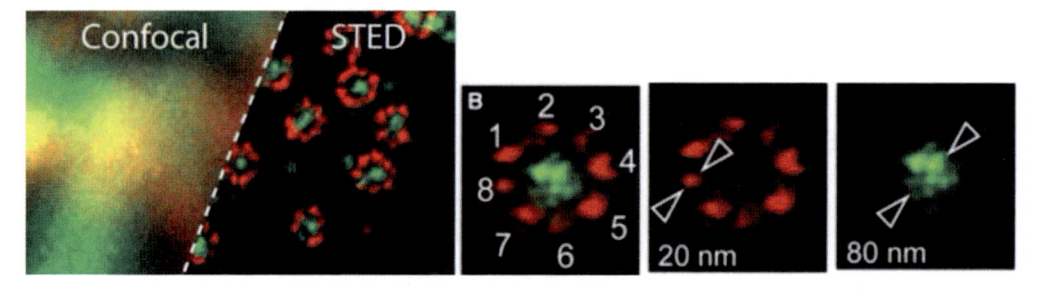

Figure 10.19 – Comparison between images taken of the Nuclear membrane with Confocal and STED microscopes [Image credit: (Göttfert, 2013)].

The best resolution of 200 nm afforded by the confocal microscope is reduced to ~20 nm by STED technique, thereby revealing details that could not have been seen through confocal fluorescence imaging.

To summarize, in Table 10.1 we list the lateral resolutions for the optical instruments we have considered in this chapter, and also included a reference to a simple microscope (single lens) built by the first person to report microbes, Antonie van Leeuwenhoek. Note the very short focal length, and the resolution of $1.35\ \mu m$. This was an amazing achievement for a time in the late 17th century. Note also that the resolution hadn't been all that much improved by the 1990s. Optical resolution using visible light still followed Abbe formula with a resolution limit of about 200 nm. The invention of STED microsopy for fluorescence imaging in 1994 clearly caused a major disruption by causing the lateral resolution to fall below 20 nm.

Year Invented	Instrument	Lateral Resolution (μm)	Reference
N/A	Human Eye	75 (typical)	Sec.10.1
~1250	Drug Store Magnifier	19 (for $m_\theta = 4$, typical)	Sec.10.2
~1670	Antonie van Leeuwenhoek Magnifier	1.35, $m_\theta = 266$, $f_{eff} \sim 0.94 mm$	van Zulen (1981)
	Compound Microscope	0.2 (theory; visible light)	Sec.10.3
1980s	TIRF Microscope	0.2	Sec.10.5
1961	Confocal Microscope	0.2	Sec.10.6
1994	STED Microscope	< 0.01 (fluorescence)	Sec.10.7

Table 10.1 – Comparision of the resolutions of optical instruments discussed in this chapter.

10.8 – Chapter Ten Exercises

Exercise 10.1 – Given that the resolving angle of the cat's eye is 1 arc minute, what is the corresponding pupil size, if this resolution is due entirely to diffraction? Hint: Treat the cat's eye as having a slit for its pupil and choose a nominal wavelength of $0.5 \times 10^{-6} m$.

Exercise 10.2 – Joe has a problem. The focal length of the lens (including cornea) in his eye is longer than his eye's length. In particular, although his eye's length is the usual 2.5 cm, his eye's lens has a focal length of 2.75 cm. Prescribe a lens in Diopters that Joe can use to correct his vision for distant objects. Give your answer in Diopters.

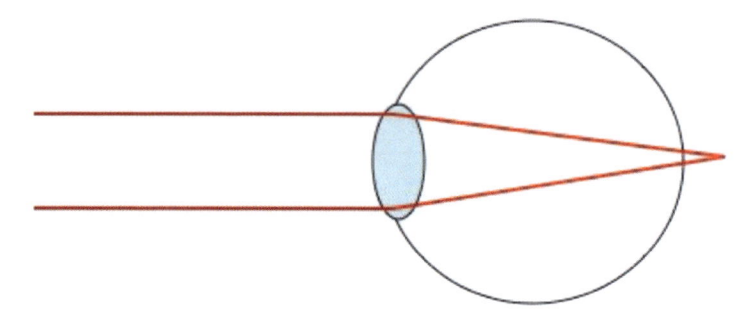

Exercise 10.3 – Total Internal Reflection Fluorescence Microscopy (TIRFM) uses evanescent fields to excite fluorescent molecules in a sample. The fluoresced light is then imaged by an objective lens, as shown in the figure below. The use of the evanescent field has the advantage of increasing axial resolution by exciting only a thin layer of the sample.

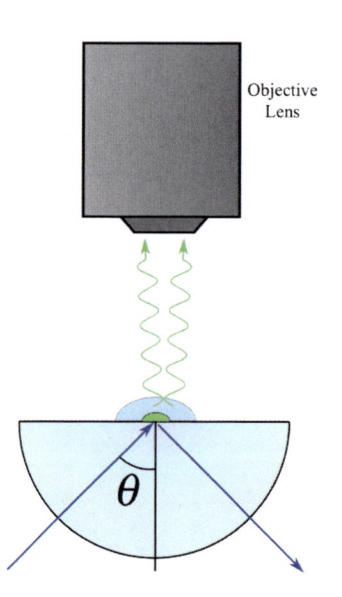

(a) A laser beam with wavelength $\lambda_0 = 458\,nm$ is directed into the semi-circular glass prism with refractive index $n_g = 1.6$, shown in the figure above, and internally reflected against the flat face at an angle of incidence $\theta = 60°$. A water droplet above the prism with refractive index $n_w = 4/3$ contains fluorescent dye molecules. Calculate the thickness of the water droplet that is made to fluoresce assuming that this thickness corresponds to the height from the interface for the intensity to drop off by a factor of e.

(b) If the fluorescent molecules emit green light with an average wavelength $\lambda_{0,g} = 532\,nm$ what amount of energy was lost to molecular vibrations for each of the blue photons that excited the sample? Recall: $\hbar = h/2\pi = 1.05 \times 10^{-34}\,J\cdot s$.

Exercise 10.4 – Antonie van Leeuwenhoek (1632-1723) discovered microbes not with a compound microscope, but with a single lens. In some cases he didn't have to grind glass to make this lens, rather he could let it form naturally. Glass by its nature is amorphous, like liquid water it flows when melted, so that a droplet of molten glass will form a spherical shape just as a water droplet does when isolated, due to surface tension. The problem with a spherical lens is the aberration caused by non-paraxial rays. To get around this Leeuwenhoek apertured the sphere (as shown below), thereby using only "paraxial" rays. Some of these lenses had a thickness (diameter) of 2 millimeter with an open aperture diameter of 0.6 millimeter, and a refractive index of 1.5. From your knowledge of geometrical optics
(a) Calculate the effective focal length of this magnifier;
(b) Calculate the position of the microbe relative to the front surface of the lens required to produce an image with a transverse magnification 200 times in comparison to the microbe;
(c) Finally, from your knowledge of microscope diffraction theory, calculate the resolution of this "simple microscope" for an illuminating wavelength of 500 nm.

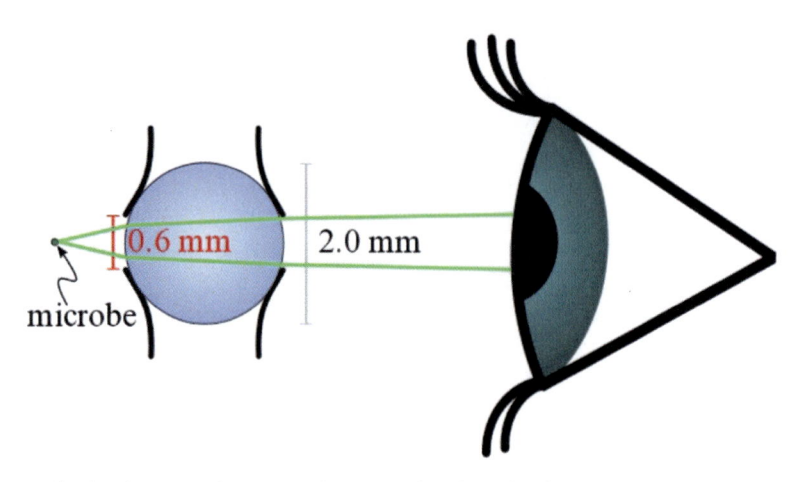

Provide analytical equations and numerical calculations with units for each of your answers. **Hint:** This simple microscope is not a "thin" lens magnifier.

Experiment #9 Smartphone microscope

This experiment allows you to use a short focal length lens to turn your smartphone into a microscope. The lens is the one which is used to collimate the light from the laser in your Pocket Optics kit. This lens is removed from the laser by unscrewing it from the laser body. Be careful to put the spring you find in a place where you can relocate it for re-assembly in the future. Now, in order to get familiar with the lens put it directly on your smartphone screen with the knurled Nickel plated end against the screen. Screw the lens containing cylinder into this Ni plated end as far as you can. Make sure that the notches on one end of the black screw mount are opposite the Ni plated knurled end (as shown below). You will notice that pixels making up the screen are easily seen even though their centers are only about 75 microns apart (see Fig.10.6). So you have generated a simple microscope of the sort that was constructed by Leeuwenhoek (see Ex.10.4). I estimate that the focal length of your lens is a bit under 7mm, and that it has a useful aperture diameter of 5mm. Given these facts you should be able to estimate the resolution of your microscope for red light (~650nm). You will notice that on one of the corners of your optical bench Abbe's resolution equation is on display. To see what might be done for a biological object, I imaged the end of a bee's foot using my iPad for illumination (see below) with my iPhone in the place of my eye. You can clearly see its articulated structure. As a measure of length, the known size of my iPad pixel taken in a separate image was compared with the bee's foot image. As you can see the sub-mm dimensions of the foot are easily discerned, and much smaller fibers are easily resolved.

Compare images taken with and without the added lens. You should be able to see small things (e.g. microbes) that cannot possibly be resolved without the additional lens. Estimate the resolution of your Smartphone microscope from the images, and compare your estimate to theory.

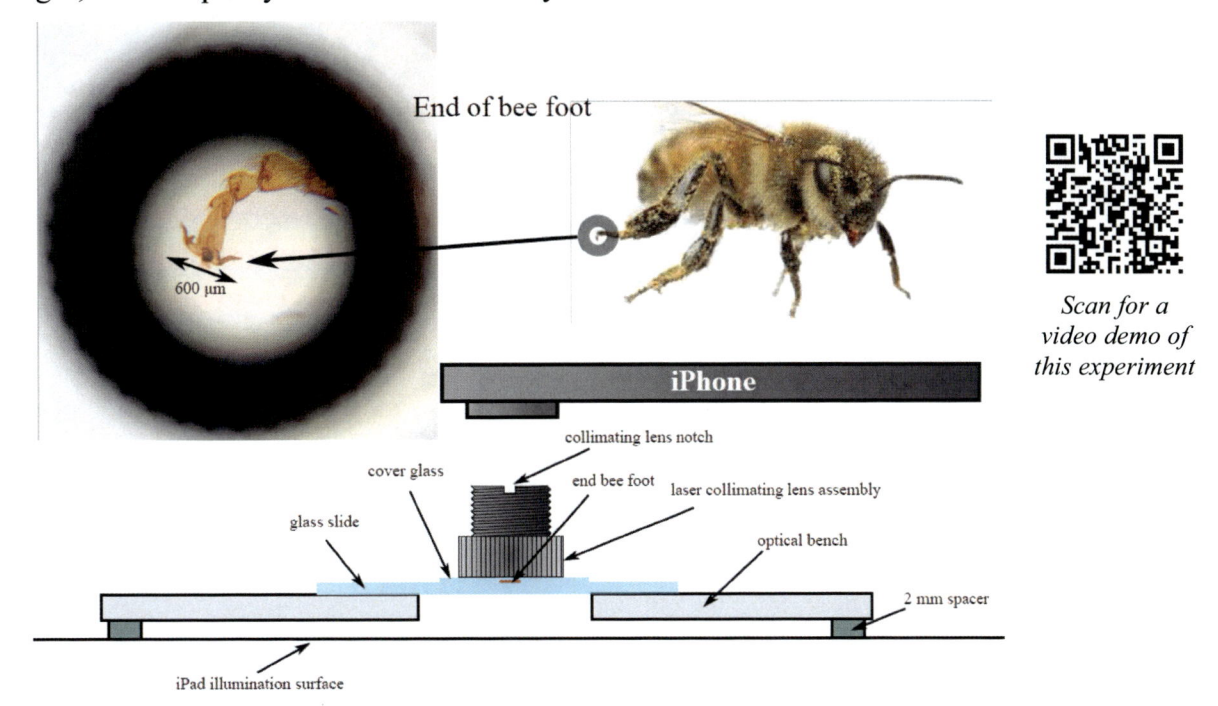

Scan for a video demo of this experiment

Appendix

A.1 – Sums and Series

In this book we use a couple of different series identities in derivations such as

$$\frac{1}{1-x} = \sum_{n=0}^{\infty} x^n = 1 + x + x^2 + x^3 + \dots . \qquad (A.1.1)$$

This series was used in discussing the transmission of an etalon in Chapter 6. In this section we discuss how such a series is generated, and give other examples.

Although we do not work with the Taylor series expansion directly it is important to note that all of the series identities in this section can be derived directly from it. The Taylor series expansion for a function $f(x)$ about point a is

$$f(x;a) = f(a) + f'(a)(x-a) + \frac{f''(a)}{2!}(x-a)^2 + \dots + \frac{f^{(n)}(a)}{n!}(x-a)^n + \dots, \qquad (A.1.2)$$

or in summation notation,

$$f(x;a) = \sum_{n=0}^{\infty} \frac{f^{(n)}(a)}{n!}(x-a)^n . \qquad (A.1.3)$$

A simple application of the Taylor series involves computing square roots of numbers up to 100 in your head. If someone were to ask you for $\sqrt{26}$ then you would naturally think that this is close to 5, the square root of $a = 25$. The function $f(x) = \sqrt{x}$, so that for $a = 25$, $f(25) = \sqrt{25} = 5$. The first order correction requires evaluating the derivative of $\sqrt{x}$ at $a = 25$, and multiplying by $(x-a)$, as instructed by Eq. (A.1.2). The derivative is

$$f'(25) = \frac{1}{2\sqrt{25}} = 0.1,$$

and $f(26) \simeq 5 + 0.1(26-25) = 5.1$, which is pretty close to the square root of 26, 5.0990.. The error here is about 20 parts in 100,000! Richard Feynman was famous for doing such computations in his head while working in the Manhattan Project as Los Alamos.

In many derivations in optics we use approximations for trig and exponential functions. The expansions for sin(x) and cos(x) about $x = 0$ are given by

$$\sin(x) = x - \frac{x^3}{3!} + \frac{x^5}{5!} + \ldots + (-1)^n \frac{x^{2n+1}}{(2n+1)!} + \ldots, \text{ and} \quad (A.1.4)$$

$$\cos(x) = 1 - \frac{x^2}{2!} + \frac{x^4}{4!} + \ldots + (-1)^n \frac{x^{2n}}{(2n)!} + \ldots. \quad (A.1.5)$$

The expansion for an exponential function e^x about $x = 0$ is

$$e^x = 1 + x + \frac{x^2}{2!} + \frac{x^3}{3!} + \ldots + \frac{x^n}{n!} + \ldots. \quad (A.1.6)$$

The expansion for the natural log of a number near 1 is

$$\ln(1+x) = x - \frac{x^2}{2} + \frac{x^3}{3} + \ldots + (-1)^{n-1} \frac{x^n}{n} + \ldots. \quad (A.1.7)$$

This expansion is valid for $|x| < 1$.

Another useful series is that for $(1+x)^n$ expanded around $x = 0$,

$$(1+x)^n = \sum_{r=1}^{\infty} \frac{n!}{r!(n-r)!} x^r = 1 + nx + \frac{n(n-1)}{2} x^2 + \ldots \quad (A.1.8)$$

It applies for $|x| < 1$, and was used in our consideration of Fraunhofer diffraction from a slit. The geometrical problem is depicted in Fig. 2.9.

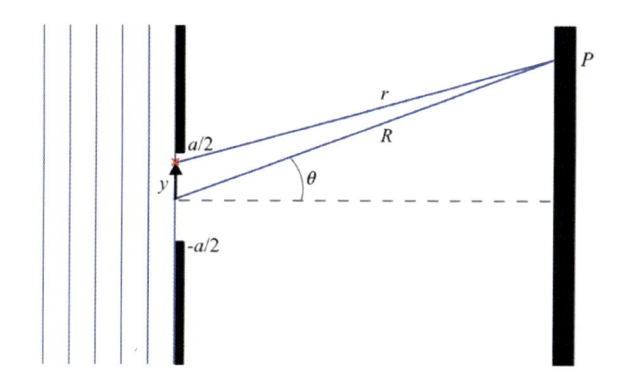

As you will recall we wanted to know how r is related y for a given angle θ, in the limit $y/R \ll 1$. We begin by deriving an equation for r in terms of R, y, and θ using the law of cosines;

$$r = \sqrt{R^2 + y^2 - 2Ry\sin(\theta)} = R\left[1 + \left(\frac{y}{R}\right)^2 - 2\left(\frac{y}{R}\right)\sin(\theta)\right]^{1/2}. \quad (A.1.9)$$

Now we express this equation in powers of y/R by using the series in Eq. A.1.8,

235

$$r \approx R + \frac{R}{2}\left[\left(\frac{y}{R}\right)^2 - 2\left(\frac{y}{R}\right)\sin(\theta)\right] - \frac{R}{8}\left[\left(\frac{y}{R}\right)^2 - 2\left(\frac{y}{R}\right)\sin(\theta)\right]^2 . \quad (A.1.10)$$

Note that we have only held onto the linear and quadratic terms in terms of y/R. By accounting for these terms only and cleaning up the expression a bit we find that

$$r \approx R\left[1 - \frac{y}{R}\sin(\theta) + \frac{1}{2}\frac{y^2}{R^2}\cos^2(\theta)\right] = R - y\sin(\theta) + \frac{1}{2}\frac{y^2}{R}\cos^2(\theta) . \quad (A.1.11)$$

The second term on the right $-y\sin(\theta)$ when multiplied by light's spatial frequency k establishes that the Fraunhofer diffraction field corresponds to the Fourier Transform of the aperture function (see Section 2.5). What deviates from that linear dependence in y is the third term. This third quadratic term when multiplied by the wave vector is a measure of how non parallel contributions add up to the amplitude on the screen. As an example, at $\theta = 0$ where Fraunhofer diffraction would ideally have every source point at a distance R from the screen, the third term provides an additional distance of $\frac{1}{2}\frac{y^2}{R}$. However the inverse R depencence allows us to reduce its effect by simply moving the screen further away. An imperative for being in the Fraunhofer region is that this additional term at the greatest y (i.e. $a/2$ in Fig.2.9) be a small fraction of the wavelength. This estabilishes a reasonable back of the envelope limit R_{Fd} for achieving a Fraunhofer diffraction image,

$$\frac{1}{2}\frac{(a/2)^2}{R_{Fd}} << \lambda \text{ ; or } R_{Fd} >> \frac{a^2}{8\lambda} .$$

A good rule of thumb often quoted is that the Fraunhofer region is for $R_{Fd} > a^2/\lambda$.

Appendix A.2 – Complex Representation of Electromagnetic Waves
Part A – Euler's Equation

Euler's equation makes a clear statement

$$e^{i\theta} = \cos\theta + i\sin\theta. \qquad (A.2.1)$$

where θ is a real number. This equation can easily be proven, by expanding the exponential using Eq. (A.1.6) and gathering real and imaginary parts. You will discover that the real part follows the series for $\cos(\theta)$ (Eq. (A.1.5)) and the imaginary part corresponds to the series for $\sin(\theta)$ (Eq. (A.1.4)). The value of Euler's equation lies in its compactness, the ease of computation of the exponential form, and the ability to convert a solution in exponential form into sinusoids. There is probably no better example than the solution to the dynamical equation for the harmonic oscillator in the form of a mass on a spring,

$$\frac{d^2 x}{dt^2} = -\frac{k}{m} x. \qquad (A.2.2)$$

All linear differential equations with constant coefficients can be solved by the trial solution $x = e^{pt}$. This trial solution satisfies Eq. (A.2.2), but only if $p^2 = -k/m$, which yields two imaginary values for p; $p_\pm = \pm i\sqrt{k/m}$. With these two exponential factors the solution to Eq. (A.2.2) is simply written as

$$x(t) = A_+ e^{i\sqrt{k/m}\,t} + A_- e^{-i\sqrt{k/m}\,t}. \qquad (A.2.3)$$

The next step is to fix the boundary conditions: for example suppose the mass is at rest at $x = A$ at $t = 0$ and let go. Together these boundary conditions tell us that $A_+ = A_- = A/2$, and

$$x(t) = \frac{A}{2}\left(e^{i\sqrt{k/m}\,t} + e^{-i\sqrt{k/m}\,t}\right). \qquad (A.2.4)$$

We have reached the stage of having the solution in terms of complex exponentials. To put it into more familiar sinusoidal functions we expand each of the complex exponential using Euler's equation with the result

$$x(t) = A\cos\left(\sqrt{k/m}\,t\right), \qquad (A.2.5)$$

the oscillation of a mass on a spring with frequency $\omega = \sqrt{k/m}$. The procedure used for our harmonic oscillator differential equation now can be applied to any linear differential equation having constant coefficients, and shows the power of Euler's equation.

Before proceeding further there are a couple of useful relationships that can easily be arrived at from Euler's equation. If the imaginary number i is changed to $-i$ the Euler's equation now reads

$$e^{-i\xi} = \cos\xi - i\sin\xi. \tag{A.2.6}$$

This is referred to as taking the conjugate. By adding Eq. (A.2.6) to Eq. (A.2.1) we arrive at a relationship between the $\cos\xi$ and the complex exponentials

$$\cos\xi = \frac{e^{i\xi} + e^{-i\xi}}{2}. \tag{A.2.7}$$

In addition, by subtracting Eq. (A.2.6) from Eq. (A.2.1) the relationship between the $\sin\xi$ and the complex exponentials is revealed;

$$\sin\xi = \frac{e^{i\xi} - e^{-i\xi}}{2i}. \tag{A.2.8}$$

Part B – Complex Representation of Electromagnetic Waves

The dynamical equation of an electromagnetic wave

$$\frac{\partial^2 E_y}{\partial x^2} = \mu_0 \varepsilon_0 \frac{\partial^2 E_y}{\partial t^2}, \tag{A.2.9}$$

is a 2nd order linear partial differential equation. Since the wave oscillates at any position x with a frequency ω, a simple test solution is $E_y = f(x)e^{-i\omega t}$, where we understand that only the real part is physical. Testing this in Eq. (A.2.9) we get

$$\frac{d^2 f(x)}{dx^2} + k^2 f(x) = 0, \tag{A.2.10}$$

where $k^2 = \mu_0 \varepsilon_0 \omega^2$. Now we have a 2nd order linear differential equation similar to the harmonic oscillator equation, which suggests the solution $f(x) = e^{px}$. With this test solution for $f(x)$, we find that $p^2 = -k^2$, or $p = \pm ik$. This gives the following overall solution

$$E_y = (E_+ e^{ikx} + E_- e^{-ikx})e^{-i\omega t} = E_+ e^{i(kx-\omega t)} + E_- e^{i(-kx-\omega t)}. \tag{A.2.11}$$

So waves having a common frequency propagate in either the plus or minus x direction, with the coefficients E_+ and E_- giving the strength of each respectively. Because we considered the solution to be the real part of E_y, in the end we will get sinusoids. Eq. (A.2.11) gives the two basic types of solution with respect to propagation, however before taking the real part it should be made clear that E_+ and E_- can carry arbitrary phases; $E_+ = |E_+|e^{i\varphi_+}, E_- = |E_-|e^{i\varphi_-}$. In addition the superposition of any number of waves is possible so long as they each satisfy the wave equation (Eq. (A.2.9)).

One particularly important example of the application of Euler's equation is the superposition of two coherent waves travelling in the same direction, but with different phase constants,

$$E(x,t) = E_0 \cos(kx - \omega t) + E_0 \cos(kx - \omega t + \phi) \qquad (A.2.12)$$

Experimentally this sum is seen to produce a single travelling wave, although it has a different amplitude and phase constant then either of the waves in Eq. (A.2.12). If you stick with the trig functions in Eq. (A.2.12), that is very hard to show unless you can remember or derive a particular identity,

$$\cos B + \cos C = 2\cos\left(\frac{B+C}{2}\right)\cos\left(\frac{B-C}{2}\right). \qquad (A.2.13)$$

Needless to say, knowing that you need this identity is not obvious, and deriving it is not easy. The neat thing about Euler's equation is that it is the "Swiss army knife" of trigonometry; it's one equation that you can use for arriving at just about all trigonometric identities.

Now we will try and tackle the problem of the superposition of the coherent waves expressed in Eq. (A.2.12) using the Euler approach, and see if one travelling wave emerges. First Eq. (A.2.12) is put in complex form,

$$E(x,t) = E_0 e^{i(kx - \omega t)} + E_0 e^{i(kx - \omega t + \phi)}. \qquad (A.2.14)$$

We can simplify this rather neatly by pulling the common factor $E_0 e^{i(kx-\omega t)}$ to the front

$$E(x,t) = E_0 e^{i(kx - \omega t)}\left[1 + e^{i\phi}\right]. \qquad (A.2.15)$$

Currently this is the sum of two terms both of which are complex. To reduce it to one complex term we remove a factor of $e^{i\phi/2}$ from the bracket;

$$E(x,t) = E_0 e^{i(kx-\omega t)} e^{i\phi/2} \left[e^{-i\phi/2} + e^{i\phi/2} \right]. \tag{A.2.16}$$

Using Eq. (A.2.7) we identify the bracket as $2\cos(\phi/2)$;

$$E(x,t) = 2E_0 \cos(\phi/2) e^{i(kx-\omega t+\phi/2)}. \tag{A.2.17}$$

Now we have the desired form of an amplitude times a complex exponential. We take the real part of this using Eq. (A.2.6), as agreed in the use of the complex exponential, and find

$$E(x,t) = 2E_0 \cos(\phi/2) \cos(kx - \omega t + \phi/2). \tag{A.2.18}$$

Eq. (A.2.18) is the electromagnetic wave produced by the overlapping of two waves that differed by a phase shift. You will notice that the phase of the wave is shifted relative to the individual waves that were combined and that the amplitude of the resultant wave is controlled by this phase shift. So we see that Euler's equation allows us to easily manipulate trigonometric expressions without knowing particular trigonometric identies.

There are limitations to the idea of converting trigonometric expressions to their complex counterparts, and then after manipulation expecting to get a clear answer by taking the real part. These problems occur when the expression is nonlinear. As an example suppose your interest is in finding the time average of $\cos^2(\omega t)$ over a cycle. It is easy to show that the result is ½. If however you wrote this out as $(e^{i\omega t})^2$, or $e^{i2\omega t}$, then the real part is $\cos(2\omega t)$, which in fact has a time average of 0. This disparity is caused by the fact that squaring a function is a non-linear operation. The only thing to do to get around this is to write out the $\cos^2(\omega t)$ completely

$$\cos^2(\omega t) = \left[(e^{i\omega t} + e^{-i\omega t})/2 \right]^2 = (e^{i2\omega t} + e^{-i2\omega t} + 2)/4. \tag{A.2.19}$$

The result on the right is just $(\cos(2\omega t)+1)/2$, and its time average is ½. There is another way to handle time averages when dealing with products, and that discussion is next.

Appendix A.3 – Time Averages

In optics we often deal with quantities that vary sinusoidally with time, i.e. the electric and magnetic field. In Appendix A.2 we learned how to represent these quantities as complex quantities of the form

$$\underline{E}(\underline{r},t) = E_0 \hat{z} e^{i(\underline{k}\cdot\underline{r}-\omega t+\phi)}, \qquad (A.3.1)$$

where ϕ indicates the phase of the wave relative to a chosen reference frame. When we have two fields with the same wavelength the time average of the product of the fields is modulated by their phase difference. For example, given fields $\hat{A}(t) = Ae^{i(-\omega t+\theta)}$ and $\hat{B}(t) = Be^{i(-\omega t+\phi)}$, where A and B are real and positive, the time average of the product $\hat{A}(t)\cdot\hat{B}(t)$ depends on the phases θ and ϕ of the two fields. The fields can be represented as

$$\hat{A}(t) = \tilde{A}e^{-i\omega t} \text{ and} \qquad (A.3.2)$$

$$\hat{B}(t) = \tilde{B}e^{-i\omega t}, \qquad (A.3.3)$$

where $\tilde{A} = Ae^{i\theta}$ and $\tilde{B} = Be^{i\phi}$. To calculate the time-average we integrate the product of the real parts of the fields over a period T. The dependence on the phase shift can be found by evaluating the time average directly

$$\left\langle \text{Re}\left[\hat{A}(t)\right]\cdot\text{Re}\left[\hat{B}(t)\right]\right\rangle_t = \frac{1}{T}\int_0^T A\cos(\omega t+\theta)\cdot B\cos(\omega t+\phi)\,dt. \qquad (A.3.4)$$

To make integration easier the cosine terms in Eq. (A.3.4) can be converted to exponential form

$$\left\langle \hat{A}(t)\cdot\hat{B}(t)\right\rangle_t = \frac{1}{4T}\int_0^T \left(\tilde{A}e^{-i\omega t}+\tilde{A}^*e^{i\omega t}\right)\left(\tilde{B}e^{-i\omega t}+\tilde{B}^*e^{i\omega t}\right)dt. \qquad (A.3.5)$$

Evaluating the product gives

$$\left\langle \hat{A}(t)\cdot\hat{B}(t)\right\rangle_t = \frac{1}{4T}\int_0^T \tilde{A}\tilde{B}e^{-i2\omega t}+\tilde{A}\tilde{B}^*+\tilde{A}^*\tilde{B}+\tilde{A}^*\tilde{B}^*e^{-i2\omega t}\,dt. \qquad (A.3.6)$$

The first and last terms in the above integral combine to form a cosine, for which the integral over an integer number of periods is 0, thus

$$\left\langle \hat{A}(t)\cdot\hat{B}(t)\right\rangle_t = \frac{\left(\tilde{A}\tilde{B}^*+\tilde{A}^*\tilde{B}\right)T}{4T} = \frac{\tilde{A}\tilde{B}^*+\tilde{A}^*\tilde{B}}{4}. \qquad (A.3.7)$$

The sum of the products $\tilde{A}\tilde{B}^*$ and $\tilde{A}^*\tilde{B}$ is twice the real part of $\tilde{A}\tilde{B}^*$, so

$$\left\langle \widehat{A}(t)\cdot \widehat{B}(t) \right\rangle_t = \frac{\text{Re}\left[\tilde{A}\tilde{B}^* \right]}{2}. \qquad (A.3.8)$$

If we now recall that $\tilde{A} = Ae^{i\theta}$ and $\tilde{B} = Be^{i\phi}$ then we can see the phase-shift dependence of the time-average

$$\left\langle \widehat{A}(t)\cdot \widehat{B}(t) \right\rangle_t = \frac{\text{Re}\left[Ae^{i\theta} Be^{-i\phi} \right]}{2} = \frac{AB\cos(\theta - \phi)}{2} \qquad (A.3.9)$$

This identity greatly simplifies taking time-averages of complex fields. It is worthwhile to work through the proof given above to convince yourself of the relation so that you can employ it when calculating time-averages for intensities such as the Poynting vector. For now, notice that when $\widehat{A}(t)$ and $\widehat{B}(t)$ are in phase, i.e. $\theta = \phi$, the time-average obtains its full amplitude. When $\widehat{A}(t)$ and $\widehat{B}(t)$ are 90° out of phase the time average of their product is 0.

Appendix A.4 – Fourier Transforms

Waveforms can be built from complex exponential functions. Possibly the simplest example is the construction of a wave frozen in space and equal to $\cos(x)$,

$$\cos(x;\lambda) = (1/2)e^{i\frac{2\pi x}{\lambda}} + (1/2)e^{-i\frac{2\pi x}{\lambda}}. \tag{A.4.1}$$

It is periodic, infinite in extent, has a characteristic spatial period λ, and was constructed from two complex exponentials with equal weights using Euler's equation. The first complex exponential can be represented by a vector in the first quadrant of the complex plane at an angle $+2\pi x/\lambda$ from the real axis, while the second is just below the real axis at angle $-2\pi x/\lambda$. Eq. (A.4.1) is also known as the complex Fourier series for $\cos(x)$. This can be understood by viewing the generalized Fourier series for an arbitrary function

$$f(x) = \sum_{n=-\infty}^{\infty} c_n e^{i\frac{2\pi n x}{\lambda}}. \tag{A.4.2}$$

Out of the infinite number of coefficients c_n, in the case of $\cos(x)$ only two coefficients are non-zero; $c_1 = c_{-1} = 1/2$. The complex Fourier series can represent any periodic waveform, such as the shark tooth wave function in Fig. A.4.1, and can be used to understand a description of aperiodic functions known as the Fourier transform. Before getting to the Fourier transform we need to arrive at an equation for the coefficients c_n.

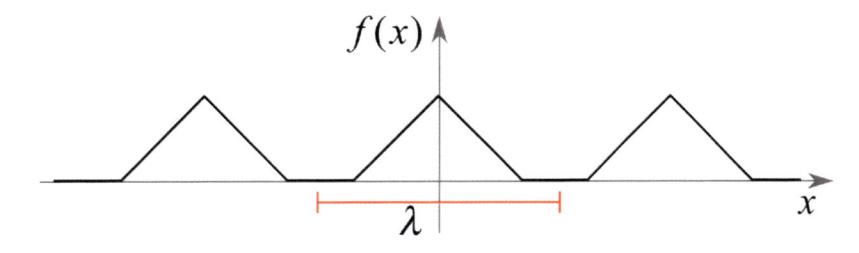

Figure A.4.1 – A shark tooth wave with period λ.

To get at c_n requires understanding a key property of the complex exponential functions. Simply stated the functions are orthonormal over their periodic length when normalized by dividing by $\sqrt{\lambda}$,

$$\int_{-\lambda/2}^{\lambda/2} \frac{e^{-i\frac{2\pi mx}{\lambda}}}{\sqrt{\lambda}} \frac{e^{i\frac{2\pi nx}{\lambda}}}{\sqrt{\lambda}} dx = \delta_{n,m}, \quad (A.4.3)$$

where $\delta_{n,m}$ is known as the Kronecker delta function, which is 1 for $n=m$ and 0 for $n \neq m$. Note that to get this result, one of the functions is conjugated. For the sake of neatness we will define the normalized complex exponential function as

$$u_n(x) = \frac{e^{i\frac{2\pi nx}{\lambda}}}{\sqrt{\lambda}}. \quad (A.4.4)$$

On this basis the ortho-normality condition reads

$$\int u_m^*(x) u_n(x) \, dx = \delta_{n,m}. \quad (A.4.5)$$

To evaluate c_n in Eq. (A.4.2) we multiply on the left by $u_m^*(x)$ and integrate over the characteristic spatial length,

$$\int u_m^*(x) f(x) \, dx = \frac{1}{\sqrt{\lambda}} \sum_{n=-\infty}^{\infty} c_n \int e^{-i\frac{2\pi mx}{\lambda}} e^{i\frac{2\pi nx}{\lambda}} dx, \quad (A.4.6)$$

from which,

$$c_m = \frac{1}{\sqrt{\lambda}} \int_{-\lambda/2}^{+\lambda/2} u_m^*(x) f(x) \, dx. \quad (A.4.7)$$

Together Eq. (A.4.2) becomes

$$f(x) = \sum_{n=-\infty}^{\infty} \left[\int_{-\lambda/2}^{+\lambda/2} u_n^*(x') f(x') \, dx' \right] u_n(x). \quad (A.4.8)$$

Let $k_n = \frac{2\pi n}{\lambda}$, then

$$f(x) = \frac{1}{\sqrt{\lambda}} \sum_{n=-\infty}^{\infty} \frac{1}{\sqrt{\lambda}} \left[\int_{-\lambda/2}^{+\lambda/2} e^{-ik_n x'} f(x') \, dx' \right] e^{ik_n x}. \quad (A.4.9)$$

If we define $\Delta k = \frac{2\pi}{\lambda} \Delta n$, where $\Delta n = 1$, then $\frac{1}{\lambda} = \frac{\Delta k}{2\pi}$, and

$$f(x) = \frac{1}{2\pi} \sum_{n=-\infty}^{\infty} \left[\int_{-\lambda/2}^{+\lambda/2} e^{-ik_n x'} f(x') \, dx' \right] e^{ik_n x} \Delta k. \quad (A.4.10)$$

Now let's suppose that the characteristic spatial period λ is increased well beyond the size of the central feature as shown in Fig. A.4.2. Since Δk is inversely proportional to λ, it will shrink and become infinitesimal as λ goes to ∞.

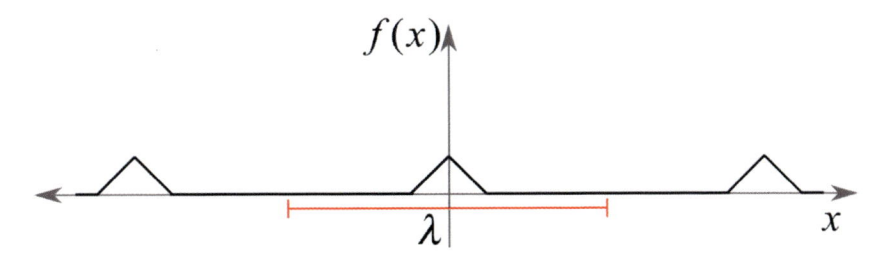

Figure A.4.2 – A shark tooth wave with period approaching infinity.

This allows the summation to turn into an integral and Δk to become differential with the result

$$f(x) = \frac{1}{2\pi} \int_{-\infty}^{+\infty} \left[\int_{-\infty}^{+\infty} e^{-ikx'} f(x') dx' \right] e^{ikx} dk . \qquad (A.4.11)$$

The integral in brackets, which we will call $g(k)$ is known as the Fourier Transform of the function f,

$$g(k) = \int_{-\infty}^{+\infty} e^{-ikx'} f(x') dx' . \qquad (A.4.12)$$

It provides information about the spectral content of the spatial function f. Once having $g(k)$, $f(x)$ can be recovered through the Inverse Fourier Transform, which from Eq. (A.4.11) is

$$f(x) = \frac{1}{2\pi} \int_{-\infty}^{+\infty} g(k) e^{ikx} dk . \qquad (A.4.13)$$

A.5 – Vector Calculus for expressing Maxwell's Eqns. as Differential Eqns.

Eqs. (1.7)-(1.10) express Maxwell's equations in differential form, and all involve the operator with the symbol ∇, which is also known as the *Del* operator. In Cartesian coordinates

$$\nabla = \hat{x}\frac{\partial}{\partial x} + \hat{y}\frac{\partial}{\partial y} + \hat{z}\frac{\partial}{\partial z}. \tag{A.5.1}$$

The simplest thing that ∇ can be applied to is a scalar function. An example of this is the application to an electrostatic potential $\varphi(x,y,z)$. For example, the electrostatic field $\underline{E} = -\nabla\varphi$. In this form $\nabla\varphi$ is referred to as the gradient of φ. The gradient of a function at a coordinate (x,y,z) is a vector in the direction of the maximum rate of change of the function. *Del* can also operate on a field vector. ZZZ

When *Del* operates on a vector field $\underline{E}$, it can do so principally in two ways. The first $\nabla \cdot \underline{E}$ produces a scalar function known as the Divergence of the field, and the second $\nabla \times \underline{E}$ produces a vector field known as the Curl of the field.

In Cartesian coordinates the Divergence of a field involves the dot product between ∇ and the field;

$$\nabla \cdot \underline{E} = \left(\hat{x}\frac{\partial}{\partial x} + \hat{y}\frac{\partial}{\partial y} + \hat{z}\frac{\partial}{\partial z}\right) \cdot (E_x\hat{x} + E_y\hat{y} + E_z\hat{z}) = \frac{\partial E_x}{\partial x} + \frac{\partial E_y}{\partial y} + \frac{\partial E_z}{\partial y}. \tag{A.5.2}$$

A field has a divergence at a point of interest, if that field is within the source of the field. This is best expressed by Maxwell's first equation; Eq. (1.7), $\nabla \cdot \underline{E} = \rho/\varepsilon_0$. Note that when the charge density vanishes at a particular location then the divergence must vanish. For an infinite sheet of charge in the y-z plane, the field must be invariant with respect to y and z, and the associated partial derivatives vanish, leaving only $\partial E_x/\partial x = 0$ just beyond the sheet; so the field outside is constant.

In Cartesian coordinates the Curl of a field involves evaluating the cross product between ∇ and the field;

$$\nabla \times \underline{E} = \left(\hat{x}\frac{\partial}{\partial x} + \hat{y}\frac{\partial}{\partial y} + \hat{z}\frac{\partial}{\partial z} \right) \times \left(E_x \hat{x} + E_y \hat{y} + E_z \hat{z} \right). \qquad (5.3)$$

There are nine terms as this expression is expanded, however the only six that are non-zero since $\hat{x} \times \hat{x} = \hat{y} \times \hat{y} = \hat{z} \times \hat{z} = 0$. The others are easily calculated by using cyclic permutations arrived at from the right hand vector rule, and represented most compactly by the circle shown below. For example $\hat{x} \times \hat{y}$ is evaluated by going clockwise around the circle for which the next unit vector is $\hat{z}$; $\hat{x} \times \hat{y} = \hat{z}$. Going counter clockwise gives the negative of the next unit vector incountered; e.g. $\hat{x} \times \hat{z} = -\hat{y}$.

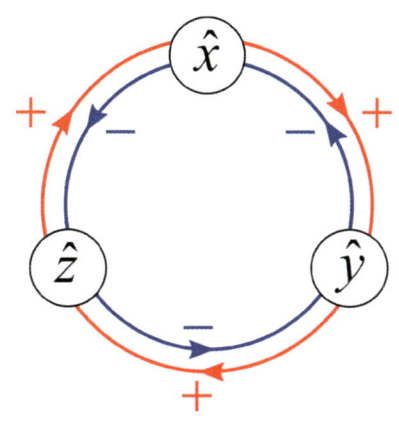

Using this cyclic permutation rule, the curl of the electric field is

$$\nabla \times \underline{E} = \hat{x}\left(\frac{\partial E_z}{\partial y} - \frac{\partial E_y}{\partial z} \right) + \hat{y}\left(\frac{\partial E_x}{\partial z} - \frac{\partial E_z}{\partial x} \right) + \hat{z}\left(\frac{\partial E_y}{\partial x} - \frac{\partial E_x}{\partial y} \right). \qquad (A.5.4)$$

As an example of the use of the curl, we return to Faraday's law, and its application to the infinite sheet of oscillating current.

In differential form Faraday's law is

$$\nabla \times \underline{E} = -\frac{\partial \underline{B}}{\partial t}. \qquad (A.5.5)$$

Our analysis of the oscillating sheet of current (Sec.1.4) established that the magnetic field is in the z-direction. So the starting point for using Faraday's law is to take the z-component of both sides of Eq. (A.5.5). In addition symmetry associated with the infinite sheet in the z-y plane requires that both the electric and magnetic fields can only be dependent on the x and time. Examining the curl of the electric field (A.5.4) in light of this symmetry shows that the z-component of the left hand side of Eq. (A.5.5) is $\dfrac{\partial E_y}{\partial x}$, so that the z-component of Eq. (A.5.5) is

$$\frac{\partial E_y}{\partial x} = -\frac{\partial B_z}{\partial t};\qquad (A.5.6)$$

precisely what was obtained from Faraday's law in integral form, Eq. (1.39).

A.6 – Take-home Experiments

To truly understand an optical phenomenon, it is necessary that it be observed, and its properties measured. This is best done as a complement to learning theory, and both should be addressed as close as possible in time. Upper level courses in Physics and Engineering seldom follow this educational duality. In addition, laboratory access should not be restricted. The Pocket Optics kit was designed to address these concerns. While at home as you are studying the material in this text book, you can test theory hands-on. For example, you can demonstrate Brewster reflection for polarizing light, question the limitations of theory, and attempt to measure the refractive index of a material. We have included ten experiments that cover many of the broader concepts discussed in the text. You are encouraged to try all of these. Better still; use the knowledge you gain from this text and your hands-on experience to create some new invention; adapt the parts you find in the kit to construct a new experiment or device. Good luck!

Experimental Report (suggested format):

Title of Experiment
Your Name

Report should include:

- a. The statement of the problem in each case (i.e. what you were asked)
- b. How you proceeded in each case
    - i. The theory that was used,
    - ii. The experimental approach where applicable (you can include diagrams and photos here).
- c. The experimental results, and their comparison with theory (where possible).
- d. Discussion and/or conclusion in each case.

Materials – The materials required for these experiments are compiled in the Pocket Optics kit produced by David Keng.

References

Cover

Arnold, S. "Microspheres, Photonic Atoms and the Physics of Nothing" *American Scientist* **89**, 414-421(2001).

Chapter One

Arnold, S. "Measuring the light infinitesimal, the theory and practice of photon counting" *Spex Speaker* **22**, 1-8 (1977).

Arnold, S., Ramjit, R., Keng, D., Kolchenko, V., and Teraoka, I. "MicroParticle Photophysics Illuminates Viral Bio-sensing" *Faraday Discuss.* **137**, 65-83 (2008).

Ashkin, A., Dziedzic, J.M., Bjorkholm, J.E., and Chu, S. "Observation of a Single-beam Gradient Force Optical Trap for Dielectric Particles" *Opt. Lett.* **11**, 288-290 (1986).

Ashkin, A., and Dziedzic, J.M. "Optical Levitation by Radiation Pressure" *Applied Physics Letters* **19**, 283-285 (1971).

Maxwell, J. C. "A Dynamical Theory of the Electromagnetic Field" *Phil. Trans. R. Soc. Lond.* **155**, 459-512 (1865).

Michelson, A. A. "Measurement of the Velocity of Light Between Mount Wilson and Mount San Antonio" *The Astrophysical Journal,* **65**, 1 (1927).

Pang, Y., H. Song, Kim, J.H., Hou, X., and Cheng, W. "Optical Trapping of Individual Human Immunodeficiency Viruses in Culture Fluid Reveals Heterogeneity with Single-molecule Resolution" *Nature Nanotechnology* **9**, 624-630 (2014).

Chapter Two

Franklin, R.E., and Gosling, R.G. "Molecular Configuration in Sodium Thymonucleate"" *Nature* **171**, 740-741 (1953).

Gabor, D. "Holography, 1948-1971" (Nobel Lecture) *Science* **177** 299-313 (1971).

Lucas, A.A., Lambin, Ph., Mairesse, R., and Mathot, M. "Revealing the Backbone Structure of B-DNA from Laser Optical Simulations of Its X-ray Diffraction Diagram" *J. Chem. Ed.* **76**, 378-385 (1999).

Watson, J. D., and Crick, F.H. "Molecular Structure of Nucleic Acids: A Structure for Deoxyribose Nucleic Acid" *Nature* **171**, 737-738 (1953).

Chapter Four

Brewster, D. "On the Laws Which Regulate the Polarization of Light by Reflection from Transparent Bodies" *Proceedings of the Royal Society of London* **2**, 14-15 (1815).

Herapath, W. B. "XXVI. On the optical properties of a newly-discovered salt of quinine, which crystalline substance possesses the power of polarizing a ray of light, like tourmaline, and at certain angles of rotation of depolarizing it, like selenite." *The London, Edinburgh, and Dublin Philosophical Magazine and Journal of Science,* **3**, 161-173 (1852).

Jones, R. C. "A New Calculus for the Treatment of Optical Systems I Description and Discussion of the Calculus" *Journal of the Optical Society of America* **31**, 488-493 (1941).

Land, E. H. "Some Aspects of the Development of Sheet Polarizers" *Journal of the Optical Society of America* **41**, 957-963 (1951).

Kahr, B, Freudenthal, J., and Kaminsky, W. "Herapathite", *Science* **324**, 1407 (2009).

Strutt, J. W. "On the Light from the Sky, Its Polarization and Colour" *Philos. Mag.* **1871**, 107-120 (1871).

Chapter Five

Brewster, D. "On the Laws Which Regulate the Polarization of Light by Reflection from Transparent Bodies" *Proceedings of the Royal Society of London* **2**, 14-15 (1815).

Chapter Six

Arnold, S., Ramjit, R., Keng, D., Kolchenko, V., and Teraoka, I. "MicroParticle Photophysics Illuminates Viral Bio-sensing" *Faraday Discuss.* **137**, 65-83 (2008).

Mcfarland, A. D., and Van Duyne, R. P. "Single Silver Nanoparticles as Real-Time Optical Sensors with Zeptomole Sensitivity" *Nano Letters* **3**, 1057-1062 (2003).

Serpenguzel, A., Arnold, S., Griffel, G. "Excitation of resonances of microspheres on an optical fiber" *Optics Letters* **20**, 654-656 (1995).

Vollmer, F., and Arnold, S. "Whispering-gallery-mode Biosensing: Label-free Detection down to Single Molecules" *Nature Methods* **5**, 591-596 (2008).

Chapter Seven

Einstein, A. "On the Quantum Theory of Radiation", *Phys. Z.* **18**, 167-183 (1917).

Maiman, T. H. "Stimulated Optical Radiation in Ruby", *Nature* **187**, 493–494 (1960).

Chapter Ten

Arpa, A., Wetzstein, G., Lanman, D., Raskar, R. "Single lens off-chip cellphone microscopy" 2012 IEEE Computer Society Conference on Computer Vision and Pattern Recognition Workshops, Providence, RI, USA (2012).

Göttfert, F., Wurm, C., Mueller, V., Berning, S., Cordes, V., Honigmann, A., and Hell, S. "Coaligned Dual-Channel STED Nanoscopy and Molecular Diffusion Analysis at 20 nm Resolution" *Biophysical Journal*, **105** (2013).

Hell, S. W. "Nanoscopy with focused light" (Nobel Lecture) *Angew. Chem. Int. Ed.* **54**, 8054–8066 (2015).

van Zuylen, J. "The microscopes of Antoni van Leeuwenhoek", *Journal of Microscopy, 121, 309-328 (1981).*

Answers to Selected Exercises

Chapter 1

Exercise 1.1 –

(a) Travelling wave with velocity $v_z = -\dfrac{\beta}{\alpha}$.

(b) Not a travelling wave.

(c) Travelling wave with velocity $v_x = -\dfrac{\beta}{\alpha}$.

(d) Travelling wave with velocity $v_x = 1$.

Exercise 1.4 – (a) $\theta = 48°$, (b) $30\,V/m$, (c) $\hat{k} = \dfrac{\sqrt{5}\hat{x} - 2\hat{y}}{3}$, (d) $\lambda = \pi \times 10^{-8}\,m$, (e) $\omega = 6 \times 10^{16}\,rad/s$, $f = \dfrac{3 \times 10^{16}}{\pi}\,Hz$, (f) $v = 3 \times 10^8\,m/s$

Exercise 1.7 – (a) $\underline{E} = (10V/m)\hat{z}\exp\left\{i\left[\dfrac{\pi}{3}\left(\dfrac{1}{\sqrt{2}}x + \dfrac{1}{\sqrt{2}}y\right) \times 10^7 - (\pi \times 10^{15})t\right]\right\}$ (b) $B_0 = \dfrac{1}{3} \times 10^{-7}\,T$, polarization: $\dfrac{(\hat{x} - \hat{y})}{\sqrt{2}}$ (c) $\langle I \rangle_t = 0.13\,W/m^2$ (d) $\underline{F} = 8.7 \times 10^{-14}\hat{x}\,N$

Exercise 1.8 – (a) $E \approx 219\,V/m$ (b) $E \approx 219 \times 10^4\,V/m$ (c) $F = 33\,pN$

Exercise 1.9 – The roots of this quadratic equation are $x = 0.115\,\mu m$ and $x = 4.460\,\mu m$. The more physically reasonable solution is $x = 0.115\,\mu m$, which indicates the scattering force pushes the virion $115\,nm$ beyond the focal point.

Exercise 1.10 – Using this value and the values supplied in the caption of Fig. 1.21

$$U_g = -7.5 \times 10^{-20}\,J = 0.47\,eV.$$

For comparison the thermal energy at room temperature ($\sim 300\,K$) is

$$E = k_B T = 4 \times 10^{-21}\,J = 0.025\,eV.$$

The energy trapping the virion is nearly 20 times the thermal energy so thermal fluctuations (Brownian motion) are unlikely to knock the virion out of the optical trap.

253

Chapter 2

Exercise 2.1 –.

(a) $\theta = 18.97^0$
(b) $\theta_{-1} = -\theta = -18.97^0$
(c) $\theta_{+1} = 77.16^0$

Exercise 2.2 – $c(k_y) = \sqrt{\dfrac{\pi}{|a|}} e^{-\dfrac{k_y^2}{4|a|}}$.

Exercise 2.3 – $I \sim 4E^2 a^2 \left[\cos(k_y D_y)\right]^2 \left[\dfrac{\sin(k_y a/2)}{k_y a/2}\right]^2$

where a is the width of a slit, and D_y is distance from the origin to the slit center.

Exercise 2.4 – $c(k_y) = \dfrac{L}{2}\operatorname{sinc}(k_y L/2) + \dfrac{L}{4}\operatorname{sinc}\left((k_g - k_y)L/2\right) + \dfrac{L}{4}\operatorname{sinc}\left((k_g + k_y)L/2\right)$.

Exercise 2.6 – $|c(k_y, k_z)|^2 = w_0^4 \pi^2 e^{-(k_y^2 + k_z^2)w_0^2/2}$, $|c(k_\rho)|^2 = w_0^4 \pi^2 e^{-w_0^2 k_\rho^2/2}$.

Exercise 2.7 – $A(z,y) = \sum_{m=0}^{5} \delta[z - a\sin(m\pi/3)]\,\delta[y - a\cos(m\pi/3)]$.

Exercise 2.8 – $c(k_z, k_y) = \sum_{m=0}^{5} e^{-i\left(k_z a \sin(m\pi/3) + k_y a \cos(m\pi/3)\right)}$.

With $Z = k_z a$, and $Y = k_y a$, the diffraction image is:

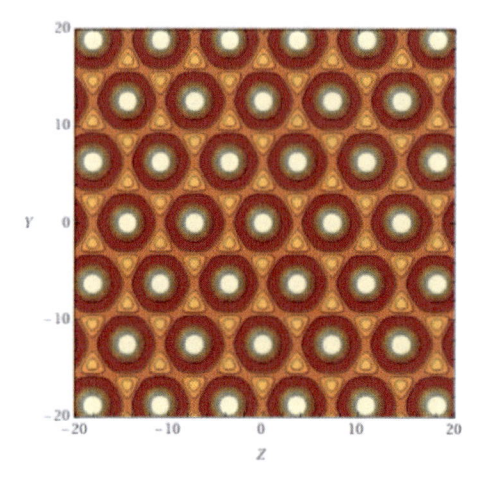

Exercise 2.10 – DNA is composed of many different atoms, the largest of which are the phosphorous atoms that form the backbone for the helix (see the figure below). Because scattering cross-section is proportional to atomic number

254

squared (Z^2) the phosphorus atoms make the most significant contribution to scattering and the other atoms can be ignored.

Exercise 2.12 – Derive $\dfrac{r}{P} = \dfrac{\cot(\alpha/2)}{2\pi}$.

Exercise 2.13 – $f = \dfrac{1}{4}$ or $f = \dfrac{3}{4}$

Exercise 2.14 – $D > 77\ nm$.

Chapter 3

Exercise 3.1 – $\dfrac{F_B}{F_E} = \dfrac{q|\underline{v} \times \underline{B}|}{q|\underline{E}|} = \dfrac{|qv_y \hat{y} \times B_z \hat{z}|}{|qE_y \hat{y}|} = \dfrac{v_y B_z}{E_y} = \dfrac{v_y}{c}$.

Exercise 3.2 – $\displaystyle\int_{-\infty}^{\infty} \sigma(\omega - \omega_0) d(\omega - \omega_0) = \dfrac{\pi |q_e|^2}{2\varepsilon_0 m_e c}$,

Exercise 3.3 – $\dfrac{\pi |q_e|^2}{2\varepsilon_0 m_e c} = 1.667 \times 10^{-5}\ m^2/s$.

Exercise 3.4 - $\sigma_{max} = 5.3 \times 10^{-20}\ m^2$.

Exercise 3.5 - $E_r = \dfrac{-|\rho_e| r}{3\varepsilon_0}$.

Exercise 3.6 – $\left(\dfrac{|q_e|^2}{16\pi^3 \varepsilon_0 m_e f_0^2} \right)^{1/3} = 1.01 \times 10^{-10}\ m$

Exercise 3.7 – $E = (\sigma_c - \sigma_{pc})/\varepsilon_0$.

Exercise 3.8 – $n^2 \approx 1 + \left(\rho_N \dfrac{|q_e|^2}{\varepsilon_0 m_e \omega_0^2} \right) \dfrac{\lambda^2}{\lambda^2 - \lambda_0^2}$.

Exercise 3.10 – $\sigma = 4.5 \times 10^{-16}\ cm^2$.

Exercise 3.11 – $I_d = I_0 e^{-\sigma \rho_N d}$

Chapter 4

Exercise 4.1 – The half-wave plate will begin to rotate about the symmetry axis.

Exercise 4.2 – $d = 858\ nm$.

Exercise 4.3 – $\underline{\underline{A}}_{LV} = \begin{bmatrix} 0 & 0 \\ 0 & 1 \end{bmatrix}$

Exercise 4.4 – $E_\theta = E_0 e^{-i\omega t} \begin{bmatrix} \cos\theta \\ \sin\theta \end{bmatrix}$

Exercise 4.5- zero transmitted

Exercise 4.7 –

(a) $\dfrac{I_0}{2}$.

(b) RCP with intensity $\dfrac{I_0}{2}$.

Chapter 5

Exercise 5.1 – $R = 0.02$,

Exercise 5.2 – The electric field of the reflected wave is phase shifted by 180° relative to the incident wave.

Exercise 5.3 – $n_1 \dfrac{\varepsilon_0 c}{2} E_i^* E_i$

Exercise 5.4 – $n^2 = \tan^2(\theta_B)$.

Exercise 5.6 – Solution: $R = 0.918$, 91.8%

Exercise 5.7 – $v_e(t) = -\dfrac{|q_e|}{m_e} E_0 \int_0^t \cos(\omega t') e^{-\gamma(t-t')} dt'$

Exercise 5.8 - $L = 39.4\,nm$.

Exercise 5.9 – $L \approx 48\ nm$

Chapter 6

Exercise 6.1 – $\tau = 1/\gamma_p$.

Exercise 6.4 – $\dfrac{I_t}{I_0} = \dfrac{1}{1 + \left[\dfrac{4R}{(1-R)^2}\left(\dfrac{nd}{c}\right)^2\right](\omega - \omega_m)^2}$

Exercise 6.5 – $E_r = \dfrac{|p_e| y_e}{3\varepsilon_0}$.

Exercise 6.6 – For etalons of certain thicknesses the phase shift caused by traversing the etalon causes destructive interference between the fields reflected from the front and back, which eliminates reflection.

Exercise 6.7 – Light trapped inside the etalon remains trapped if it can't transmit through the sides. The higher the Reflectance R the smaller the transmittance T. The linewidth increases with the rate of loss of trapped energy.

Exercise 6.10 – $|H(\omega)|^2 = \dfrac{A^2}{a^2 + (\omega_0 - \omega)^2}$

Exercise 6.11 – $|H(\omega)|^2 = \dfrac{A^2 \pi}{a} e^{-(\omega-\omega_0)^2/2a}$,

Exercise 6.12 – $\tau = \dfrac{dn}{(1-R_2)c}$, $(\delta\omega)_{1/2} = \dfrac{(1-R_2)c}{dn}$.

Exercise 6.13 – $\Rightarrow m \approx 183 (rounded)$

Exercise 6.14 – $n_{eff} \approx 1.5$

Chapter 7

Exercise 7.1 – $\mathcal{E} = 2.48 \ eV$

Exercise 7.2 – $\Delta E = 6 \times 10^{-20} \ J$

Exercise 7.4 – $\lambda = 17 \ \mu m$

Exercise 7.5 – $I(\lambda) = \dfrac{2hc^2}{\lambda^5} \dfrac{1}{e^{hc/(\lambda kT)} - 1}$

Chapter 8

Exercise 8.1 – The widths for the 0th and 1st order peaks are both $2\pi/(NP)$.

Exercise 8.2 – $I_d(\xi) = 2 + 2\cos\left[\dfrac{(k_2 - k_1)\xi}{2}\right]\cos\left[\dfrac{(k_2 + k_1)\xi}{2}\right]$.

Exercise 8.3 – $n = 1.0003$

Chapter 9

Exercise 9.1 – If you examine the limit where the light travels entirely along the x-axis but reflects off of the surface once you will see the distance is twice the semi-major axis or $2A$.

Exercise 9.3 – $f = 0.8 \ m$.

Exercise 9.6 – $f \approx 11 \ mm$.

Exercise 9.7 – $f_{eff} = \dfrac{nD}{4(n-1)}$.

Exercise 9.8 – $f = 0.125 \ m$.

Exercise 9.9 – $\begin{bmatrix} 0.92 & -4.79 m^{-1} \\ 0.033 m & 0.92 \end{bmatrix}$

Exercise 9.10 – $R_2 = -2.5 \ cm$

Exercise 9.12 – $\begin{bmatrix} 0.37931 & -0.62069 \ mm^{-1} \\ 1.37931 \ mm & 0.37931 \end{bmatrix}$

Chapter 10

Exercise 10.1 – $(\theta)_{limit} \simeq 1.7\,mm$.

Exercise 10.2 – $\dfrac{1}{f_g} = +3.6\,D$

Exercise 10.3 – (a) $L \approx 97\ nm$, (b) $\Delta E = 6 \times 10^{-20}\,J$.

Exercise 10.4 – (a) $f_{eff} = 1.5\,mm$, (b) $0.4925\,mm$ to the left of the ball lens

(c) $1.27\,\mu m$ in air, and $1.27\,\mu m / 1.33$ in water.

Index

A

Abbe's diffraction limit, 226
absorption, 87, 166
absorption cross-section, 87
Aether, 180
aluminum, 134
Ampère-Maxwell Equation, 2
amplitude, 7
antenna, 143
aperture, 51, 54, 76
aperture function, 54
attenuation, 53, 102, 163

B

Babinet's principle, 59, 60
bandwidth, 170
beam expander, 46
beam splitter, 46, 180
Beer's Law, 105
Birefringence, 114
Boltzmann constant, 165
Bragg peaks, 68
Brewster reflection, 109
Brewster's angle, 131

C

capacitance, 23, 92, 142
capacitor, 22, 91, 142
cat's eye, 206
characteristic length, 133
charge, 23
chromatic aberration, 95
circular polarizer, 120
Circularly polarized light, 114
classical atom, 83, 85
coherent, 43, 163
complex conjugate, 86
complex conjugate transpose, 119
confocal pinhole, 223
conservation of energy, 128
constructive interference, 45
Coulomb's law, 89
Crick, Francis, 68, 69
critical angle, 133
crystal, 68
crystallographer, 68
current, 2
cuvette, 102
Cy5, 105

D

damping rate, 101
delta function, 57
destructive interference, 45
detuning, 88, 103

dichroic polarizer, 111
dielectric, 28, 91, 124
dielectric constant, 92
diffraction, 43, 52, 76
diffraction grating, 46, 50
diode laser, 176
Diopter, 198
dipole, 32, 93
dipole moment, 32, 95
dissipation, 135
dissipation rate, 97
DNA, 50, 59, 67, 70
doping, 195
double helix, 72
double strand, 74
drag coefficient, 134

E

Earth, 15, 210
Einstein, Albert, 27, 165, 180, 225
electric
    field, 2
    flux, 2
electrode, 34
emf, 2
energy, 22, 166
energy density, 24, 166
etalon, 147
evanescent field, 134
experiment, 69

F

Fabry-Perot etalon, 147
Faraday's Law, 2, 125
farsightedness, 198
Fermat's least time principle, 188
finesse, 158
focal length, 192, 200, 201, 204
focus, 200
force, 28
Fourier transform, 54, 56, 61, 68, 154, 177, 181, 243
Franklin, Rosalind, 68, 69
Fraunhofer, 52
free spectral range, 158
frequency, 22
frequency domain, 153
Fresnel, 52
FTIR, 180

G

Gabor, Dennis, 50
Galileo Galilei, 15
Gallium Indium Phosphide, 172
gamma radiation, 21
Gauss's Law, 1, 90, 92, 104, 145, 149
geometrical optics, 188, 189
Germanium, 195

Gosling, Raymond, 68, 69
gradient, 33
grating, 48
gravity train, 144
group velocity, 100

H

helical structure, 68, 69
Herapath, William Bird, 112
Herapathite, 112
hologram, 50
holographic diffraction grating, 46
holographic grating, 46
holography, 50
horizontal polarizer, 120
H-sheet, 113
Hubble satellite, 190
Hubble telescope, 190
Huygen's principle, 51
hydrogen, 90

I

incandescence, 166
incident wave, 125
inductance, 142
inductor, 39, 142
infrared radiation, 22
intensity, 54, 68, 181
interference, 43, 68
interferogram, 181
Io, 15
iPhone camera, 105
iris, 50, 210
irradiate, 167

J

Jones Calculus, 117
J-sheet, 112
Jupiter, 15

K

Kirchhoff's circuit law, 126

L

Land, Edwin, 107
Land-Wheelwright Labs, 107
laser, 163
laser cavities, 155
Law of Malus, 114
layer lines, 72
Leeuwenhoek, Antonie van, 217
left circular polarized (LCP), 115, 119
lens power, 198
light, 1, 15, 21, 165, 189, 195
line width, 88, 155
linear dichroism, 111
linearly polarized, 120
linearly polarized light, 114
linewidth, 158

longitudinal mode, 170
Lorentz atom, 84, 89, 90
Lorentz force, 85
Lorentz, Hendrik Antoon, 84
Lorentzian, 88, 97, 161
LuxMeter, 105

M

magnetic
 field, 2
magnetic flux, 126
Maiman, Theodore, 168
Maxwell, James Clerk, 1
Michelson interferometer, 180
Michelson, Albert, 16, 180
Minsky, Marvin, 224
mirror, 135
molarity, 103
monochromatic light, 181
MP3L, iii, 159
Mt. San Antonio, 17
Mt. Wilson, 17
multiple scattering, 61

N

nanoparticle, 153
nearsightedness, 198
Newton' Third Law, 29
Newton's second law, 6
Nobel Prize, 50, 68, 165
nucleus, 90

O

Ole Rømer, 15
optical cross-section, 101
optical path length, 188, 189, 190
optical tweezers, 29, 30
orbit, 15
orbital, 90
oscillation, 145
oscillator, 89
oscillator strength, 101

P

parabolic mirror, 191
paraxial rays, 194
particle, 27, 154
permeability, 2
permittivity, 1, 92
phase, 240
phase shift, 74, 116
phase velocity, 100
phasor, 86
phosphate backbone of DNA, 70
Photo 51 of DNA, 69
photoelectric effect, 27
photographic film, 46, 48
photon, 26, 27, 117
photon lifetime, 155, 157
pitch, 49, 72

Planck's radiation law, 165
plane-polarized light, 121
Planetarium, 193, 196
plasma, 135, 146
plasma frequency, 135, 145
plasmonics, 146
polarizability, 32
polarization, 91, 108
polarization density, 95
polarized light, 108
Polaroid Corporation, 107
Polymeric Iodine, 113
population inversion, 168, 169
potential, 23
power, 24, 105, 201
Poynting theorem, 86
Princess Leia, 50
probability density, 90
proton, 84
pupil, 50

Q

Q, quality factor, 157
quantum
quantum mechanics, 89, 149, 164
quantum optics, 117
quantum physics, 4, 225
quarter-wave plate, 121
quinine, 112

R

radiation
    electromagnetic, 15
radiation force, 87
radiation pressure, 40, 85
radiowaves, 22
Rayleigh scattering, 108
real image space, 194
real object space, 194
reference
    frame, 8
reflection, 124
reflection coefficient, 125, 148
reflectivity, 163
refraction, 97
refractive index, 94, 95, 124, 206
relaxation rate, 87, 134, 136, 145
resolution, 224
resonance, 86
resonant mode, 171
resonator, 158
rest mass, 27
right circular polarized, 115, 118
Ruby laser, 169

S

semi-classical atom, 91
semiconductor, 146, 170
silica, 195
silver, 49

Snell's law, 81, 110, 128, 189, 195
space, 60
spatial decay, 97
spectral density, 182, 186
spectrum, 21
speed of light, 15
spherical aberration, 194
spontaneous emission, 165, 166, 226
Star Wars, 50
STED, 225
stimulated emission, 163, 165
stimulated emission depletion microscope, 225
Sun, 15
superposition, 43, 50
swimming pool, 194

T

Taylor expansion, 32, 234
thin lens, 201
time, 69, 101, 189
transmission coefficient, 125, 147
transverse wave, 4
traveling waves, 18
tuner, 143

U

Ultra-Violet, 145

V

vacuum, 1
variable capacitor, 143
velocity, 18
vertical polarizer, 120
vertically polarized, 120
virtual image, 195, 219
virtual object, 196, 200
visible spectrum, 22
voltage, 23

W

Watson, James, 68, 69
wave, 70, 240
wave equation, 43
wave packet, 100
wave vector, 19
wavelength, 7, 22
WGM, 158
Wilkins, Maurice, 68

X

X-ray diffraction, 67
X-rays, 22, 68

Y

Young's double-slit, 59

Made in the USA
Middletown, DE
12 August 2024